# CHEMICAL-MECHANICAL POLISHING OF LOW DIELECTRIC CONSTANT POLYMERS AND ORGANOSILICATE GLASSES
*Fundamental Mechanisms and Application to IC Interconnect Technology*

# CHEMICAL-MECHANICAL POLISHING OF LOW DIELECTRIC CONSTANT POLYMERS AND ORGANOSILICATE GLASSES
*Fundamental Mechanisms and Application to IC Interconnect Technology*

by

**Christopher L. Borst**
*Texas Instruments, Inc.
Dallas, TX*

**William N. Gill**
*Rensselaer Polytechnic Institute
Troy, NY*

**Ronald J. Gutmann**
*Rensselaer Polytechnic Institute
Troy, NY*

KLUWER ACADEMIC PUBLISHERS
Boston / Dordrecht / London

**Distributors for North, Central and South America:**
Kluwer Academic Publishers
101 Philip Drive
Assinippi Park
Norwell, Massachusetts 02061 USA
Telephone (781) 871-6600
Fax (781) 681-9045
E-Mail: kluwer@wkap.com

**Distributors for all other countries:**
Kluwer Academic Publishers Group
Post Office Box 322
3300 AH Dordrecht, THE NETHERLANDS
Telephone 31 786 576 000
Fax 31 786 576 254
E-Mail: services@wkap.nl

 Electronic Services < http://www.wkap.nl>

**Library of Congress Cataloging-in-Publication Data**

Borst, Christopher L. (Christopher Lyle)
 Chemical-mechanical polishing of low dielectric constant polymers and organosilicate glasses : fundamental mechanisms and application to IC interconnect technology / by Christopher L. Borst, William N. Gill, Ronald J. Gutmann
  p. cm.
 Includes bibliographical references and index.
 ISBN 1 4020 7193 0
  1. Interconnects (Integrated circuit technology) 2. Semiconductors—Polishing.
 3. Grinding and Polishing I. Gill, William N. II Gutmann, Ronald J. III Title.

TK7874.53 .B67 2002
621.3815'2—dc21

2002028782

Copyright © 2002 by Kluwer Academic Publishers

All rights reserved. No part of this work may be reproduced, stored in a retrieval system, or transmitted in any form or by any means, electronic, mechanical, photocopying, microfilming, recording, or otherwise, without the written permission from the Publisher, with the exception of any material supplied specifically for the purpose of being entered and executed on a computer system, for exclusive use by the purchaser of the work.

Permission for books published in Europe: permissions@wkap.nl
Permission for books published in the United States of America: permissions@wkap.com

*Printed on acid-free paper.*
Printed in the United States of America.

# TABLE OF CONTENTS

Preface ............................................................................ xi

Acknowledgements ...................................................... xiii

Chapter 1: Overview of IC Interconnects ............................ 1
    1.1    SILICON IC BEOL TECHNOLOGY TRENDS ........... 2
    1.2    SIA ROADMAP INTERCONNECT
           PROJECTIONS ................................................... 6
    1.3    LOW-κ REQUIREMENTS AND MATERIALS .......... 7
           1.3.1    Fluorinated Glasses ........................................ 8
           1.3.2    Spin-On Glasses and Silsesquioxanes .............. 8
           1.3.3    Organosilicate Glasses ................................... 9
           1.3.4    Polymers ...................................................... 9
           1.3.5    Fluorinated Polymers ................................... 11
           1.3.6    Porous Media .............................................. 11
    1.4    NEED FOR LOW-κ CMP PROCESS
           UNDERSTANDING ............................................ 12
    1.5    SUMMARY ........................................................ 14
    1.6    REFERENCES .................................................... 14

Chapter 2: Low-κ Interlevel Dielectrics .............................. 17
    2.1    FLUORINATED GLASSES ................................. 19
    2.2    SILSESQUIOXANES .......................................... 21
    2.3    ORGANOSILICATE GLASSES ........................... 24
    2.4    POLYMERS ....................................................... 27

| | | |
|---|---|---|
| 2.5 | FLUORINATED HYDROCARBONS | 34 |
| 2.6 | NANOPOROUS SILICA FILMS | 35 |
| 2.7 | OTHER POROUS MATERIALS | 38 |
| 2.8 | REFERENCES | 40 |

**Chapter 3: Chemical-Mechanical Planarization (CMP)** ............ 45

| | | |
|---|---|---|
| 3.1 | CMP PROCESS DESCRIPTION | 47 |
| 3.2 | CMP PROCESSES WITH COPPER METALLIZATION | 50 |
| | 3.2.1 Oxide CMP | 50 |
| | 3.2.2 Copper CMP | 50 |
| | 3.2.3 Damascene Patterning | 52 |
| | 3.2.4 CMP Targeted Results and Challenges | 55 |
| 3.3 | CMP OF LOW-κ MATERIALS | 55 |
| 3.4 | CMP PROCESS MODELS | 62 |
| | 3.4.1 Contact Mechanics-Based Models | 63 |
| | 3.4.2 Fluid Mechanics-Based Models | 64 |
| 3.5 | LANGMUIR-HINSHELWOOD SURFACE KINETICS IN CMP MODELING | 66 |
| 3.6 | REFERENCES | 71 |

**Chapter 4: CMP of BCB and SiLK Polymers** ............ 71

| | | |
|---|---|---|
| 4.1 | REMOVAL RATE IN COPPER SLURRIES | 71 |
| 4.2 | SURFACE ROUGHNESS | 74 |
| 4.3 | SURFACE AND BULK FILM CHEMISTRY | 78 |
| | 4.3.1 Angle-Resolved Surface Results | 78 |
| | 4.3.2 Depth Profiling Results | 84 |
| 4.4 | EFFECT OF CURE CONDITIONS ON BCB AND SILK REMOVAL | 85 |
| | 4.4.1 Variation in Cure Conditions | 86 |
| | 4.4.2 Effect of Cure Conditions on Removal Rate | 86 |
| 4.5 | EFFECT OF CMP AND BCB AND SILK FILM HARDNESS | 88 |

# TABLE OF CONTENTS

4.6 COMPARISON OF BCB AND SILK CMP
WITH OTHER POLYMER CMP .......................... 91

4.7 SUMMARY ................................................. 93

4.8 REFERENCES ............................................. 94

## Chapter 5: CMP of Organosilicate Glasses .......................... 97

5.1 EFFECT OF FILM CARBON CONTENT ................. 97
    5.1.1 Removal Rate in Oxide Slurries .................... 98
    5.1.2 Removal Rate in Copper Slurries ................. 101

5.2 SURFACE ROUGHNESS ................................. 103

5.3 SURFACE AND BULK FILM CHEMISTRY .......... 106
    5.3.1 XPS Surface Results ................................. 107
    5.3.2 FTIR Bulk Profiling Results ....................... 109

5.4 COPPER DAMASCENE PATTERNING
WITH OSG DIELECTRICS ................................ 110
    5.4.1 Hardmasks or Dielectric Cap Layers ............ 111
    5.4.2 Low-κ Etching ....................................... 112
    5.4.3 CMP Integration ..................................... 113

5.5 SUMMARY ................................................ 117

5.6 REFERENCES ............................................ 117

## Chapter 6: Low-κ CMP Model Based on Surface Kinetics ........... 119

6.1 ISOLATION OF THE CHEMICAL
EFFECTS IN SILK CMP ................................. 120

6.2 CMP WITH SIMPLIFIED "MODEL"
SILK SLURRIES .......................................... 122
    6.2.1 Removal Rate Dependence on Slurry
Reactant Concentration ............................. 124
    6.2.2 KH Phthalate Slurry for Copper
CMP Applications ................................... 127
    6.2.3 SiLK Removal Rate Dependence
on Velocity ........................................... 128

6.3 PHENOMENOLOGICAL MODEL FOR
CMP REMOVAL ........................................... 129
    6.3.1 Applicability to the CMP of BCB and SiLK ..... 131
    6.3.2 Applicability to the CMP of Organosilicate
Glasses ................................................ 133

- 6.4 FIVE STEP REMOVAL MODEL USING MODIFIED LANGMUIR-HINSHELWOOD KINETICS FOR SILK CMP ..... 134
  - 6.4.1 Five Step Surface Mechanism ..... 135
  - 6.4.2 Implementation into 3-D Fluid Mechanics and Mass Transport Models ..... 139
  - 6.4.3 Results ..... 141
- 6.5 TWO STEP REMOVAL MODEL USING HETEROGENEOUS CATALYSIS FOR SILK CMP ..... 152
  - 6.5.1 Two Step Surface Mechanism ..... 152
  - 6.5.2 Results ..... 154
- 6.6 EXTENDIBILITY OF MODEL TO DESCRIBE THE CMP OF OTHER MATERIALS ..... 155
  - 6.6.1 Copper ..... 156
  - 6.6.2 Dielectrics ..... 156
- 6.7 REFERENCES ..... 158

## Chapter 7: Copper CMP Model Based Upon Fluid Mechanics And Surface Kinetics ..... 161

- 7.1 FLOW MODEL ..... 163
- 7.2 COPPER REMOVAL MODEL ..... 164
  - 7.2.1 Mass Transport of Oxidizer to the Wafer Surface ..... 165
  - 7.2.2 Kinetic Surface Steps ..... 166
  - 7.2.3 Copper Removal Rate and Effectiveness Factor ..... 168
  - 7.2.4 Kinetic Rate Parameters ..... 170
  - 7.2.5 Solution Procedure ..... 171
- 7.3 MODEL RESULTS ..... 172
- 7.4 COPPER CMP EXPERIMENTS WITH POTASSIUM DICHROMATE BASED SLURRY ..... 176
- 7.5 SUMMARY ..... 179
- 7.6 REFERENCES ..... 179

## Chapter 8: Future Directions in IC Interconnects and Related Low-κ ILD Planarization Issues ..... 181

- 8.1 PLANARIZATION OF INTERCONNECTS WITH ULTRA LOW-κ ILDS ..... 182

|  |  |  |
|---|---|---|
| | 8.1.1 | Alternatives to Conventional CMP .............. 182 |
| | 8.1.2 | Non-CMP Approaches to Planarization ......... 183 |

8.2 ALTERNATIVES FOR THE POST-COPPER/ULTRA LOW-κ INTERCONNECT ERA ........................ 185
    8.2.1 Alternative Materials ............................... 185
    8.2.2 Alternative Distribution for Signals and Clocks ................................................. 186
    8.2.3 Non-electrical Distribution of Signals and Clocks ................................................. 188
    8.2.4 Non-Planar Integrated Assemblies – Three Dimensional (3D) Integration ..................... 189

8.3 3D WAFER-SCALE INTEGRATION USING DIELECTRIC BONDING GLUES AND INTER-WAFER INTERCONNECTION WITH COPPER DAMASCENE PATTERNING ................ 193
    8.3.1 Wafer Bonding ....................................... 194
    8.3.2 Inter-wafer Interconnect ........................... 197
    8.3.3 Comparison with Other Wafer-Scale 3D Integration Technologies ........................... 199

8.4 SUMMARY AND CONCLUSIONS ...................... 199

8.5 REFERENCES .............................................. 200

## APPENDICES

Appendix A: EXPERIMENTAL PROCEDURES AND TECHNIQUES ....................................... 203

Appendix B: XPS DEPTH-PROFILE DATA ........................ 221

Appendix C: CMP DATA FOR ANOMALOUS SILK REMOVAL BEHAVIOR ........................................ 225

Index ................................................................. 227

# PREFACE

As semiconductor manufacturers implement copper conductors in advanced interconnect schemes, research and development efforts shift toward the selection of an insulator that can take maximum advantage of the lower power and faster signal propagation allowed by copper interconnects. One of the main challenges to integrating a low-dielectric constant (low-κ) insulator as a replacement for silicon dioxide is the behavior of such materials during the chemical-mechanical planarization (CMP) process used in Damascene patterning. Low-κ dielectrics tend to be softer and less chemically reactive than silicon dioxide, providing significant challenges to successful removal and planarization of such materials.

The focus of this book is to merge the complex CMP models and mechanisms that have evolved in the past decade with recent experimental results with copper and low-κ CMP to develop a comprehensive mechanism for low- and high- removal rate processes. The result is a more in-depth look into the fundamental reaction kinetics that alters, selectively consumes, and ultimately planarizes a multi-material structure during Damascene patterning. Chapter one presents an overview of IC interconnects, focuses on the increased complexity and process development in the last decade, and discusses the technology progression towards copper and low-k dielectrics.

Chapters two and three provide a more detailed review of the two main subjects of the book -- low-dielectric constant materials and the CMP process. Chapter two discusses the properties of low-κ films that are desired for interconnect performance, listing characteristics and deposition methods for films ranging from silicon dioxide to polymers and organosilicate glasses (OSGs) to porous media. Chapter three introduces the CMP unit process and the fundamental goals of polishing/planarization, as well as the various approaches that have been used to mathematically model the complex mechanisms for film removal. The two subjects overlap in the latter sections of Chapter three in a summary of recent results for the CMP removal of low-κ films.

Chapters four and five discuss specific work in the CMP of three candidates for near-term Cu/low-κ integration, benzocyclobutene (BCB), SiLK, and OSGs. Chapter four compares different polishing characteristics observed between BCB and SiLK, in terms of removal rate, physical alteration, and chemical alteration. While BCB has a glass transition temperature (Tg) lower than desired for ICs, the chemical composition and property differences compared to SiLK provides a useful contrast in establishing a fundamental understanding of the CMP process. Chapter five does the same for OSG, in an effort to understand the differences and similarities in removal mechanism between the two families of dielectric being implemented at the k=2.7 technology node. In addition, the challenges and rewards involved in the integration of OSG and Cu into a Damascene-patterned interconnect are described in Chapter five.

Chapter six combines experimental observations and known CMP mechanisms to present a comprehensive mechanism that includes aspects of the surface kinetics, fluid mechanics, mass transport and shear-enhanced removal that occur during CMP polishing of low-κ dielectrics. The discussion ranges from phenomenological discussion of atomic-scale surface mechanisms to numerical calculations of wafer-scale removal rate of SiLK polymer. Chapter seven extends and broadens the numerical CMP model to further understand the mechanisms for Cu CMP removal. In both chapters full wafer CMP data is presented that is agreement with model predictions.

The last chapter, Chapter eight, extends the scope of interconnect processing to reflect our perspective on future materials and planarization approaches such as ultra low-κ dielectrics (very porous dielectrics and air gap structures), low-stress planarization techniques, and three-dimensional integration of planar ICs through wafer bonding.

At the Rensselaer Polytechnic Institute Center for Integrated Electronics and Electronics Manufacturing, we have been involved in CMP-related research for more than a decade, often in collaboration with leaders of the high-technology IC industry. The many research programs in copper, oxide, and low-κ CMP have provided the knowledge base and opportunities for fundamental learning that are central to this book. Our approach is meant to uncover the fundamental aspects of a complex process that continues to be central to the advancement of copper interconnect technology.

C. L. Borst  
W. N. Gill  
R. J. Gutmann

# ACKNOWLEDGEMENTS

The authors gratefully acknowledge the financial support of the American semiconductor industry in funding our research in copper interconnect technology at Rensselaer for over a decade. In particular, support by the Semiconductor Research Corporation (SRC), Dow Chemical Company, Texas Instruments, IBM, Intel, AMD, Applied Materials, IPEC/Planar, and Lam/OnTrak has been critical in the Rensselaer interconnect research program in copper metallization, chemical-mechanical polishing (CMP), low dielectric constant (low-$\kappa$) materials, and CMP modeling.

This book is the direct result of collaboration between the Rensselaer Center for Integrated Electronics and Electronics Manufacturing and the Rensselaer Department of Chemical Engineering, and includes work performed under direct support by the SRC, Texas Instruments and Intel Corporation.

The authors acknowledge Rensselaer faculty colleagues for their contributions to and discussions involving copper and low-$\kappa$ CMP modeling, including T.S. Cale, D.J. Duquette, S.P. Murarka, and D. Schwendeman. Additional acknowledgement for research interactions at Rensselaer is due to T. Apple, J. Crivello, and M. Tomozawa. Interactions with industrial and other university researchers have also been extremely helpful, particularly R. Geer, V. Korthuis, L. Jiang, J.D. Luttmer, M. Rutten, J. Ryan, E. Schaffer, S. Shankar, G. Shinn, and M. Simmonds. The contributions of many Rensselaer graduate students, group members, and cleanroom staff had a positive impact on this work. Dr. Dipto, G. Thakurta, presently at Intel, had a particularly strong impact on our modeling work. Chapter 7 on the CMP on copper is based upon his PhD thesis and related published work.

The authors would like to gratefully acknowledge the many authors whose contributions to the literature serve as the basis for the background information provided, and the many journals and book publishers who allowed the work to be used in this book. Many thanks to Ms. Betty Lawson for compiling and formatting the manuscript, and to A. Jain and S. Ponoth for providing references for low-$\kappa$ materials.

Most importantly, the authors would like to thank our families for providing the support that made this book possible.

C.L.B
W.N.G
R.J.G

# Chapter 1

# OVERVIEW OF IC INTERCONNECTS

For many years integrated circuit (IC) interconnect structures were relatively simple, as the main IC processing concern was fabrication of higher density transistors. During these initial two decades of IC development (early 60s to early 80s), the back-end-of-the-line (BEOL) interconnect structure was neither a performance limiter nor a manufacturing cost enhancer. As front-end-of-the-line (FEOL) technology improved with lithography and other advances allowing more-and-more transistors to be incorporated on an IC chip, the interconnect structure grew in fabrication complexity to become a significant component of manufacturing cost (mid 80s). Moreover, since the interconnect performance does not benefit from dimensional scaling as do the transistors in an IC, the interconnects became a performance factor in high speed digital ICs, due to the resistance-capacitance (R-C) delay.

By the late 80s, these trends were well recognized, and the interconnect era of IC technology was launched. The impact of the interconnect structure on IC performance and cost was increasingly recognized, and BEOL research became as important as continued FEOL research. The interconnect technology decade of the 90s included four major technology advances with significant impact on BEOL manufacturing processes and IC performance capabilities. These four technology advances were:

1) chemical-mechanical planarization (CMP), which enables six-eight levels of on-chip metallization with high manufacturing yield and acceptable cost
2) copper (Cu) metallization for lines and vias to replace aluminum (Al) lines and tungsten (W) vias for improved electrical resistance and electromigration capability
3) dual damascene (DD) patterning to replace metal reactive ion etching (RIE) and dielectric gap fill to provide improved line definition and lower BEOL manufacturing cost
4) low dielectric constant (low-κ) interlevel dielectrics (ILDs) to replace $SiO_2$ for lower line and coupling capacitance, combined with copper metallization to reduce R-C delay.

CMP has been well established in IC manufacturing in the past decade, although the fundamental principles of this most rapidly growing IC unit process are not well established. Copper metallization and DD patterning have been introduced in many IC products and will clearly be fully incorporated in microprocessors, application-specific ICs (ASICs) and high-performance system-on-a-chip (SOC) implementations in all leading-edge manufacturing lines. Low-$\kappa$ ILDs with CMP copper metallization and DD patterning are just being incorporated in products today, with dominant incorporation in advanced microprocessors, ASICs and SOCs projected in the next three to five years.

## 1.1 SILICON IC BEOL TECHNOLOGY TRENDS

The four main BEOL innovations that affect interconnect-limited Si ICs (CMP, copper metallization, DD patterning, and low-$\kappa$ ILDs) have been the result of intense research and development efforts to reduce the impact of on-chip interconnection on both electrical performance and yield/manufacturing cost. Performance and cost become increasingly important as the IC minimum feature size (MFS) has reduced from 750 to 110 nm in the past decade, with 50-nm technology a current research and development priority. Even with full implementation of these four innovations within the next few years, on-chip interconnection will remain a key design issue in high-performance microprocessors, ASICs, and SOCs [1.1].

Global planarization using CMP was the first of these innovations, developed by IBM and used in IC products first by IBM, Intel, and Micron in the early 90s. Currently all major IC manufacturers use CMP for microprocessors and ASICs, where four to six levels of metallization are required to achieve minimum chip area [1.2, 1.3]. The planarization provided by CMP allows an almost unlimited stacking of metallization levels in ICs, constrained only by real-estate requirements of conducting vias for interlevel interconnection (and to the Si devices) and by yield considerations. CMP has been the most rapidly growing IC unit process in the latter half of the 90s with approximately a 30% compound annual growth rate (CAGR) for equipment and consumables (mainly slurries and pads).

Copper metallization has been actively pursued in many research laboratories for over a decade, since copper has improved electrical conductivity and electromigration resistance [1.4-1.7]. The IBM announcement in September 1997 that copper would be used in the next generation of advanced IC chips (with such chips presently going into products) has intensified copper metallization efforts at major IC companies. The IBM six-level copper interconnect (demonstrated with oxide ILD and W local interconnects) has superior electrical performance and is claimed to result in a higher yield, lower cost manufacturing process. The latter feature, if demonstrated by other manufacturers, will result in widespread implementation of copper metallization even where the performance advantages are not needed (similar to the increasing use of CMP in Si IC manufacturing plans with only three levels of metallization).

With copper metallization and oxide ILDs, diffusion barriers and adhesion promoters are necessary to prevent copper penetration into $SiO_2$. Copper is known to diffuse rapidly under the high electric-field strengths (~1 MV/cm) within the ILDs and to degrade electronic transport properties (e.g. to increase minority-carrier lifetime in Si). TaN, Ta and TiN are currently the barriers of choice, deposited along with a seed layer of Cu by physical vapor deposition (PVD) or by CVD. Following the PVD (or CVD) of liner and seed, Cu is electroplated to fill vias and trenches. Electrochemical deposition (ECD) had not been previously used in Si IC manufacturing, but research and development efforts with copper, leveraged to prior developments in first-level packaging, quickly indicated very promising results [1.8, 1.9].

Incorporation of copper for IC metallization requires a new patterning strategy. In conventional ICs, patterning is achieved by RIE of aluminum lines, followed by deposition of an ILD, patterning of the ILD for vias, CVD of W (or, in less demanding applications, PVD of Al), and CMP of W to result in a planarized dielectric surface with W-vias. After PVD of Al for the next interconnect level, the process is repeated. Unfortunately (in the opinion of many), RIE of copper has not been possible without elevated processing temperature, which requires a hard mask (i.e. a material used for transferring a photoresist image when the photoresist can not maintain its structure in subsequent processing) and increasing process complexity. As a result, the process described for the W vias -- damascene (or inlaid metal) patterning -- has been incorporated for both lines and vias.

For further reduction in process complexity, the DD process has been developed in which both lines and vias are formed in the same process, requiring two lithography steps but one copper-deposition and one CMP step. While the damascene process is another BEOL IC manufacturing process change, development of a high-yield DD process provides reduced manufacturing cost and improved line width control, particularly with the different interconnect dimensions used for six or more metallization levels. A three-level fully-planarized interconnect structure is depicted in Figure 1.1, where metal level 1 is a single damascene patterned line and metal levels 2 and 3 are DD patterned lines and vias.

Various DD patterning strategies have been investigated in order to obtain a robust manufacturing process. Four alternatives have been described for oxide ILDs, with two additional processes developed for low-κ ILDs using clustered hard masks [1.11-1.13]. These alternatives, summarized in Figure 1.2, involve tradeoffs in various unit process steps, particularly lithography and ILD etch. While all have relative advantages, the via-first deep etch is the most utilized among the four traditional techniques and the use of clustered hard masks (or some modification) is advantageous with polymer ILDS [1.13]. Damascene patterning is more fully described in Chapter 3.

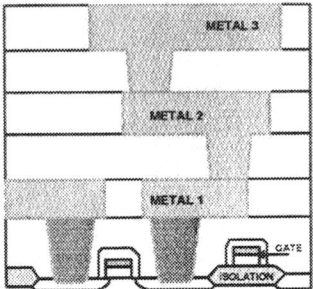

*Figure 1.1* Schematic of damascene-patterned, three-metal level interconnect structure [1.10]

The fourth BEOL innovation is the use of low-κ materials for ILDs to reduce propagation delays and reduce electrostatic coupling between lines. Coupling between adjacent lines on the same interconnect level or parallel lines on adjacent levels in close proximity can reduce interconnect performance. While the performance advantages of low-κ are greater than

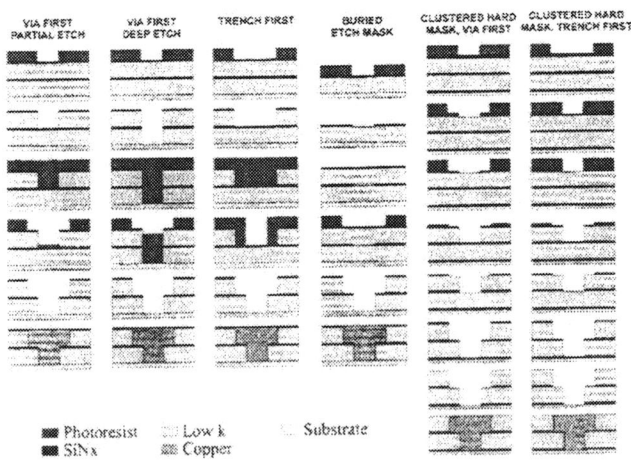

*Figure 1.2* DD patterning strategies for copper/low-κ dielectrics. The first four are conventional approaches developed for oxide dielectrics, while the two clustered hard mask techniques established for polymer dielectrics (after [1.13]).

for Cu in many designs and less process changes are involved in incorporating low-κ materials, the incorporation of low-κ is following the introduction of copper in IC manufacturing. One reason is the large number of choices of low-κ material [1.14-1.16] and the desire to have a material which can be used for at least three generations of technology; a second reason is the difference in process requirements for the ILD with RIE-patterned metal and DD patterning. With copper already in products, and with well-accepted manufacturing cost advantages of Cu with DD patterning, the industry is focusing on low-κ with Cu as a near-term development issue. Due to the number of low-κ alternatives and the need to establish a manufacturing base with Cu and DD, the incorporation of low-κ ILDs in products will be introduced at the κ=2.7 node in 2003, with widespread use by 2005.

Consider the six-level copper interconnect structure (with one level of local W interconnect and W vias for device contact) with single-level damascene patterning at the first copper level (M1 level) and DD patterning for M2-M6, as depicted in Figure 1.3(b). The larger line widths at the upper levels is typical in ICs with multiple interconnect levels. While the IBM technology depicted uses conventional $SiO_2$ as the ILD throughout, the incorporation of low-κ materials reduces parasitic electrostatic coupling.

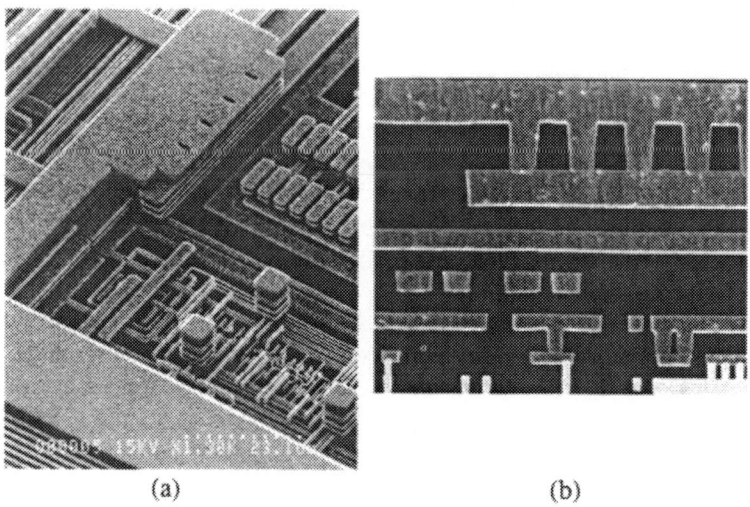

*Figure 1.3* Six-level damascene patterned interconnect structure as reported by IBM (from IBM web page). Note that the local interconnect level and vias to the device contacts are tungsten, the first copper level (M1) is single-damascene patterned and levels M2-M6 are DD patterned. (a) Three-dimensional perspective after etching of oxide dielectric. (b) Focused ion beam-scanning electron microscopy (FIB-SEM) cross section.

For this reason, IBM has selected "silicon-applications low-k" dielectric, or SiLK™ as a leading low-κ ILD option. SiLK integration is discussed further in Chapters 3, 4 and 6.

## 1.2 SIA ROADMAP INTERCONNECT PROJECTIONS

The Semiconductor Industry Association (SIA) has coordinated industry-wide projections of semiconductor technology and classified issues to guide research and development programs for the past 20 years. On-chip interconnect was one of the paramount roadmap issues during the interconnect decade of the 90s [1.17], and will be discussed here.

The Interconnect section of the 1997 National Technology Roadmap for Semiconductors (NTRS) indicates the following five difficult challenges which must be addressed before 2006 with minimum feature size (MFS) of 100 nm and above:

1) chip reliability issues with copper, damascene patterning, and low-κ ILDs
2) process integration with Cu/low-κ and Al/low-κ, particularly low cost, high yield, and acceptable reliability
3) barriers that solve integrations issues with Cu/low-κ
4) dimensional control of feature sizes and film thickness to control interconnect electrical impact (and the required metrology)
5) low-κ materials with the electrical, mechanical, and thermal requirements for high-performance ICs

Table 1.1 is a snap-shot of the 1997 NTRS interconnect technology requirements, reflecting challenges in research and development. Since the original publishing of this table, development efforts have kept pace with the roadmap, exploring several dielectric options with $3.5>\kappa>2.7$ and $2.7>\kappa>2.2$. However, the multiple options available have delayed manufacturing implementation.

While recent results indicate that there are no show stoppers for DD-patterned copper interconnects with oxide ILDs at the 130 nm technology node, the low-κ ILD alternatives to date have various deficiencies. The low-κ selection will continue to be a process integration technology challenge, particularly for $\kappa<2.2$. These ultra low-κ ILDs with Cu are being delayed by a combination of the need to gather manufacturing experience with copper and DD patterning as well as the aforementioned process integration issues.

™ SiLK is a trademark of The Dow Chemical Company

## OVERVIEW OF IC INTERCONNECTS

Table 1.1 Projected microprocessor interconnect technology requirements.

| YEAR<br>TECHNOLOGY<br>NODE | 1999<br><br>180 nm | 2000 | 2001 | 2002<br><br>130 nm | 2003 | 2004 | 2005<br><br>100 nm |
|---|---|---|---|---|---|---|---|
| Gate length (nm) | 230 | 210 | 180 | 160 | 145 | 130 | 115 |
| Number of Metal Levels | 6-7 | 6-7 | 7 | 7-8 | 8 | 8 | 8-9 |
| Conductor Effective Resistivity (μΩ-cm) Cu wiring | 2.2 | 2.2 | 2.2 | 2.2 | 2.2 | 2.2 | 2.2 |
| Barrier Thickness (nm) | 17 | 16 | 14 | 13 | 12 | 11 | 10 |
| Interlevel Metal Insulator Effective Dielectric Constant (κ) | 3.5-4.0 | 3.5-4.0 | 2.7-3.5 | 2.7-3.5 | 2.2-2.7 | 2.2-2.7 | 1.6-2.2 |

Solutions exist
Solutions being pursued
No known solution

The five difficult challenges delineated for beyond 2006, with MFS below 100 nm include the following:
1) dimensional control issues, including metrology, *in situ* process control, and computer-aided design (CAD) techniques
2) aspect ratios for metal fill and ILD etch, particularly with DD patterning
3) barriers which need to be thin to fully utilize Cu and low-κ ILDs;
4) FEOL-benign processes, that is BEOL process which do not damage scaled devices (e.g., soft etch and clean)
5) solutions after Cu/low-κ, such as optical interconnects, multiplexed interconnects and three-dimensional (3D) integration (as discussed in Chapter 8)

Clearly, the role of interconnect-limited design in Si IC requires additional interconnect modeling so that the complicated BEOL interconnect structure can be accounted for in the same way as FEOL device design. Such BEOL design tools are in a stage of relative infancy.

## 1.3 LOW-κ REQUIREMENTS AND MATERIALS

At this time low dielectric constant alternatives to silicon dioxide glass are being introduced into IC manufacturing. The near term low dielectric constant target is the 2.5 - 3.0 range, which includes polymer materials and organic/inorganic hybrids. As mentioned in the ITRS, the subsequent low-κ 2.0 - 2.2 range implies either the incorporation of porosity into one of the known dielectrics or introducing a fluorinated polymer, both of which present challenges. Porous materials require extensive experimental evaluation, some of which are discussed in subsequent chapters (mostly 2, 4 and 5) in this book. Such an evaluation examines the effective dielectric constant, water absorption, chemical reactivity, temperature stability, and compatibility of each material with copper metal and barrier materials. A more complete description of dielectrics of interest is presented in Chapter 2.

### 1.3.1 Fluorinated Glasses

Fluorinated silicon dioxide glass (FSG) was the first low-κ dielectric implemented in industrial manufacturing. FSG is similar in structure and properties to common silicon dioxide dielectric. FSG is deposited [1.18], [1.19] using plasma-enhanced chemical vapor deposition (PECVD) to obtain films with κ ranging from 3.5 – 3.7 [1.20]. During deposition, fluorine becomes substituted in the $SiO_2$ structure, either at bonding sites or interstitial locations. The result is a dielectric film that can be "tuned" by the deposition parameters to contain a desired amount of fluorine, film hardness, and dielectric constant. One concern with FSG in practical integration is the reactivity of the fluorine within the film. The unbound fluorine can migrate to the dielectric interface and react with conductor [1.21] or barrier materials, and the Si-F bond itself may react with water in the FSG film [1.22].

### 1.3.2 Spin-on Glasses and Silsesquioxanes

Spin-on glasses (SOG) are formed by the liquid-phase polymerization of methyl- and methoxy- substituted silane $[(CH_3)_x(CH_3O)_ySiH_z$, where x + y + z = 4] monomers [1.23]. The films are deposited on the wafer surface by spin-coating, then heated to initiate the polymerization reaction. The reaction forms a polymer structure, the extent of which depends on the cure time and temperature. The dielectric constant of these materials typically ranges from 2.9 – 3.6, depending on the amount of substitution of hydrogen and methyl groups in the $SiO_2$ structure [1.24]. Hydrogen silsesquioxane (HSQ) and methyl silsesquioxane (MSQ) are specific types of SOG that are the most physically and chemically stable due to the formation of linked cage bonding structures. The main concerns with HSQ, MSQ, and SOG in general are (1) variations in material properties (strength, κ, water content) with cure conditions, and (2) decomposition of methyl or hydrogen during heating or plasma treatment [1.25]. Each of these issues is a challenge for use in a successful interconnect scheme.

### 1.3.3 Organosilicate Glasses

Organosilicate glasses (OSG) are silicon dioxide films doped with carbon to lower dielectric constant. Carbon is generally incorporated in the $SiO_2$ structure during the chemical vapor deposition (CVD) process by replacing the silane reactant ($SiH_4$) with a methyl silane (($CH_3)_xSiH_y$). During the plasma deposition, the methyl groups become randomly bound to silicon atoms. Doping a film with carbon terminates some of the silicon bonds within the oxide lattice, leading to a reduction in film density and a reduced number of electrical dipoles present in the film per unit volume [1.26]. The result is lower density carbon-doped oxide film with a κ of 2.7 – 3.0. OSG dielectric constant is tunable as a function of the amount of carbon incorporated into the film and the film density. Both of these parameters are a function of the reactant concentration and the plasma power used during deposition [1.27].

OSG are very desirable low-κ candidates due to their similarity to oxides, their ability to be deposited by CVD, and the lack of mobile fluorine that may react with interconnect metals [1.27]. OSG also have very high glass transition temperature (~ 600 °C), and low moisture absorption due to their hydrophobic character. The main difficulty with OSG is that the films contain both oxide-like and carbon-like content, creating challenges in etch and CMP selectivity. Etch chemistries that are designed to remove photoresist will attack the carbon in the OSG structure, leaving behind silanol (Si-OH) groups which increase the dielectric constant.

### 1.3.4 Polymers

In general, polymers have low dielectric constants (2.0 – 3.0) due to the nature of their bonding and structure. There is a wide variety of organic polymer dielectrics under consideration, all deposited using either CVD or spin-on polymerization. Both methods result in long carbon bonded chains with low oxygen content. The main contributor to the stability of organic dielectrics is the six-membered benzene ring, which is one of the most stable forms of carbon bonding.

Organic materials can be tailored using synthetic chemistry to be flexible and hydrophobic, so as to absorb negligible moisture that could reduce their insulating ability. Due to the high temperatures involved in semiconductor processing, successful dielectric candidates must have high temperature stability. This is a significant challenge for polymers. The glass transition temperature, $T_g$, separates a solid structure from a supercooled liquid one and is an important property of polymers.

Polyimides [1.28] are linear polymers that contain carbon, oxygen, and nitrogen. The presence of double bonded oxygen in the structure degrades electrical performance due to water absorption and higher dielectric constant. The polyimides do not form crosslinked bonds and are thus called "thermoplastics". Thermoplastics, in general, have lower film strength and lower glass transition temperature than densely crosslinked thermosets. Polyimides are well-known to synthetic chemists and can be tailored using different monomers to have a variety of characteristics such as chain length and oxygen content.

Poly(arylene) ethers [1.29] are named for their *aryl*, or benzene structures, connected by single oxygen atoms, or *ether* linkages. They have high temperature stability due high content of benzene rings. The oxygen linkage provides flexibility in the chain to prevent thin film cracking and allow maximum crosslinking. Due to low oxygen content, poly(arylene) ethers have low dielectric constant and low water uptake.

Parylene [1.30] materials are chemical vapor deposited materials that form long linear chains. The fact that the materials may be deposited by CVD makes them very versatile for filling small dimensions. Their structure consists solely of carbon and hydrogen, allowing for high temperature stability and very low water uptake. The main disadvantage of parylene is its lack of crosslinking, resulting in poor film strength.

DVS-BCB [1.31], or divinyl siloxane bis-benzocyclobutene, is a hybrid polymer that contains a small amount of inorganic silicon along with organic carbon, oxygen, and hydrogen. BCB is a spin on polymer that reacts upon curing to form a highly crosslinked thermoset structure. This results in good mechanical strength and thermal stability. BCB polymer also exhibits excellent barrier properties for hindering copper diffusion through the polymer dielectric [1.32].

Aromatic hydrocarbons such as SiLK [1.33] are polymer materials that consist almost completely of benzene rings. This structure provides the highest temperature stability possible for organic materials, and makes the structure relatively inert to chemical attack. The lack of oxygen or nitrogen makes the polymers hydrophobic, with very low water uptake. These materials are prepared by spin-coating and curing monomer solutions that crosslink to form the polymer films.

The main integration concerns regarding low-κ polymers are their lower temperature stability, thermal conductivity, and mechanical stability. Currently, the controlled collapse chip connection (C4) technique used in flip chip packaging requires the interconnect to withstand temperatures above 400 °C [1.34], which is above the $T_g$ for several of the polymers listed in Table 1.3. Polymer films do not conduct heat away from the interconnect as well as $SiO_2$, and are also mechanically weaker due to their amorphous, carbon-based structure. However, polymers are finding a great deal of interest in the industry due to their low dielectric constant, ease of deposition, and ability to reduce interconnect stress.

### 1.3.5 Fluorinated Polymers

The class of dielectric that has the best insulating properties (lowest $\kappa$) without incorporated porosity is fluorinated polymers. Fluorine addition, as in the case of fluorinated oxides, terminates chains within the polymer material with very chemically and electrically inert bonds. This reduces the polarizability of the dielectric, and thus increases its effectiveness as an insulator. Techniques have been used to fabricate fluorinated polyimides with $\kappa \sim 2.5$ [1.35] and derivatives of Teflon®AF with $\kappa$ as low as 1.9 [1.36]. Fluorinated polymers have integration problems due to adhesion issues and diffusion/migration of fluorine from the polymer to metal surfaces [1.37].

### 1.3.6 Porous Media

Several research and development efforts have been made to introduce air into existing dielectric materials to produce porous materials with $\kappa \sim 1.5$. Porous silica [1.38] has been demonstrated to have a tunable $\kappa$ (depending on the porosity) of between 1.5 and 2.5. The porous silica "xerogels" are made from sol-gel processes that require acid/base catalysis. Porous organic materials have also been made using two-phase polymer systems, of which one phase is selectively volatilized following deposition, leaving the structure porous. This method has been used to generate nanoporous methyl silsesquioxanes with $\kappa$ as low as 2.2 [1.39, 1.40]. Recent efforts have been focused on the development of a porous crosslinked polymer such as SiLK. This material will have a $\kappa$ in the range of 1.5 – 2.0, depending on porosity. In general, porous options have low dielectric constant but also very low fracture toughness and adhesion strength. The porosity leads to problems in fabrication and integration that must be addressed before these materials can be seriously considered manufacturable [1.41].

---

® - Teflon is a Registered Trademark of E.I. du Pont de Nemours and Company

## 1.4 NEED FOR LOW-κ CMP PROCESS UNDERSTANDING

Ideally a low-κ ILD is not in direct contact with the slurry in a damascene or dual-damascene patterning process. The conducting liner, a Ta/TaN sandwich structure in today's mainstream copper interconnect technology, prevents direct contact during the first-step of a copper CMP process. During the subsequent second-step CMP process of removing the conducting liner, a hard inorganic dielectric capping layer is typically used to provide mechanical support and prevent interaction between the slurry and the low-κ material. However, as the minimum feature size goes below 100 nm in the near future, the conducting liner and dielectric cap become performance limiters since a minimum thickness is needed to remain effective. Clearly new approaches are desired, minimizing the thickness of both liners and caps, while trying to select materials and operating parameters which might eliminate liners and caps entirely.

A full discussion of the need for conducting liners as adhesion promoters and diffusion barriers (APDBs) is beyond the scope of this book. However, the need that is well established for copper and oxide ILDs [1.2, 1.6, 1.11, 1.12] is different with low-κ ILDs. Particularly with polymer ILDs, a conventional adhesion promoter may not be needed and diffusion of copper under high values of electric field is often less than with oxide [1.42-1.44]. Moreover, small amounts of an alloying element such as magnesium has been shown to form an in-situ barrier for copper, with an interfacial layer formed in post-deposition annealing [1.45-1.47]. These research results indicate that a copper interconnect strategy may indeed be feasible without a conducting liner requirement; unfortunately, approaches to eliminate the need for a dielectric cap layer over low-κ materials have not been as fully explored yet.

Interaction of copper slurries with low-κ ILDs is required to move toward an advanced copper interconnect patterning strategy and is the focus of this book. Particular emphasis is placed on the understanding of materials suitable for the κ = 2.7 node, announced for product manufacturing as early as the end of 2002 and projected for use through 2004 in the ITRS Roadmap shown in Table 1.1. However, more recent projections indicate that this 1999 ITRS projection may not be achieved and that κ = 2.7 ILDs may be incorporated in leading fabs for a longer period of time (see Chapter 8 for our projections of the long-term interconnect future).

Irrespective of the timetable, CMP process interactions with low-κ materials need to be understood in order to:
1) evaluate scaling limitations of copper interconnect technology by knowing the impact of direct slurry interaction with dielectrics

2) explore material and process parameters that seem to be the most promising approach to eliminate the need for conventional barriers and low-κ dielectric caps
3) assist in delineating paradigm shifts in technology necessary to alleviate the interconnect bottleneck after κ < 2.1, a no-known-solution area of the ITRS (as discussed in Chapter 8)

The approach taken is to fully evaluate the CMP of low-κ materials (specifically κ = 2.7) in copper (and conducting liner) slurries. Our emphasis is on material removal rate and post-CMP surface condition (roughness, chemical composition, scratch density and post-CMP surface cleanliness). Clearly a slurry which removes copper, conducting liner, and ILD at controlled rates without surface and bulk material changes is highly desirable.

In addition to such an experimental evaluation of the materials of most current interest for the κ = 2.7 technology node, fundamentally based models for CMP of low-κ materials is presented (Chapters 6 and 7). While a two-step phenomenological model is often applicable in CMP, a five-step model using Langmuir-Hinshelwood kinetics (Chapter 6) is established to provide fundamental insight into polymer CMP. The polymer of most interest at κ = 2.7, SiLK, is used in quantifying the model parameters.

Our low-κ CMP work has clearly established the ability to polish many low-κ materials without severely damaging the films. Previously, CMP of low-κ polymers was unresolved and even fundamentally questioned as polymers are often chemically inert and mechanically fragile. With appropriate slurry chemistry and reasonable CMP tool parameters, removal rates of manufacturing interest can be obtained without appreciable surface scratching, with low surface roughness, and without appreciable post-CMP surface modification. These results provide a data base for enhanced understanding of these and other materials and for encouraging the extension of novel approaches with oxide ILDs (such as in-situ barrier formation without dielectric caps) to be pursued with low-κ ILDs. The knowledge base may indeed provide new processes for other integration applications as well, including integrated optics with organic waveguides, microelectromechanical systems (MEMS) and micro-fluidic devices.

While low-κ CMP processes have been demonstrated, the integration of ultra low-κ ILDs with copper metallization is proving to be very difficult. Incorporation of low-κ and ultra low-κ ILDs is one area of the 2001 ITRS that is extended in time compared to the 1999 ITRS. In comparison, most technology factors, such as minimum features size (MFS), have been condensed in time.

## 1.5 SUMMARY

This chapter presents an overview of this book. Initially, a perspective on IC interconnects was presented, highlighting the 90s and the transition to (1) full interconnect planarity enabling an increasing number of interconnect levels and establishing an industrial infrastructure for CMP, (2) copper metallization for improved line resistance and current-handling ability (i.e., improved electromigration), (3) dual damascene patterning required by copper and leading to simplified BEOL equipment sets and lower manufacturing cost, and (4) low-κ ILDs to reduce R-C delay and coupling capacitance.

Following a review of near-term NTRS (1997) and ITRS (1999 and 2001) projections, low-κ ILD requirements were summarized and various materials leading from κ = 3.9 for $SiO_2$ to < 2.0 for porous dielectrics are briefly summarized (a more thorough treatment is presented in Chapter 2). The chapter concludes with a discussion of the need for low-κ CMP understanding, both in providing an information base for scaling of IC interconnects and in guiding further research.

## 1.6 REFERENCES

[1.1] National Technology Roadmap for Semiconductors and International Technology Roadmap for Semiconductors: San Jose, CA: Semiconductor Industry Assoc., 1997, 1999 and 2001.

[1.2] J.E. Steigerwald, S. P. Murarka, and R.J. Gutmann, *Chemical-Mechanical Planarization of Microelectronic Materials*, New York: Wiley, 1997.

[1.3] Chemical-Mechanical Planarization ULSI Multilevel Interconnection (CMP-MIC) Annual Conf. Dig., 1996-2001.

[1.4] D. Edelstein et al., Int. Electron Devices Meeting, 1997, pp. 773-776

[1.5] C.K. Hu and J.M.E. Harper, *Materials, Chemistry, Phys.*, **52**, 6 (1998).

[1.6] S.P. Murarka and R.J. Gutmann, Eds., Copper Metallization for Future VLSI, *Materials, Chemistry, Phys.* (Special Issue), **41**, 159 (1995).

[1.7] T.A. Alford, J. Li, J.W. Mayer, and S.-Q. Wang, Eds., *Copper-Based Metallization and Interconnects, Thin Solid Films* (Special Issue), **262**, 1 (1995).

[1.8] P.C. Andricacos, C. Uzoh, J.O. Dukovic, J. Korkans, and H. Deligianni, *IBM J.*, **42**, 557 (1998).

[1.9] P.C. Andricacos, *Interface*, **7**, 23 (1998).

[1.10] S. Wolf and R.N. Tauber, Silicon Processing for the VLSI Era, 2$^{nd}$ Edition, CA: Lattice Press (1999).

[1.11] R.J. Gutmann et al., *Thin Solid Films*, **270**, 472 (1995).

[1.12] D.T. Price, R.J. Gutmann, and S.P. Murarka, *Thin Solid Films*, **308-309**, 523 (1997).

[1.13] D.T. Price and R.J. Gutmann, presented at the 1998 Advanced Metallization Conf.

[1.14] W.W. Lee and P.S. Ho, *Mater. Res. Bull.*, **22**, 19 (1997).

[1.15] R.J. Gutmann et al., in *Proc. Materials Res. Soc.*, **381**, 177 (1995).

[1.16] Dielectrics for ULSI Multilevel Interconnection (DUMIC) Annual Conf. Dig., 1995-2001.

[1.17] R.J. Gutmann, *IEEE Trans. Microwave Theory Tech.*, **29**, 667 (1999).

[1.18] S.-M. Lee, M. Park, K.-C. Park, J.-T. Bark, J. Jang, Jpn. *J. Appl. Phys., Part 1* **35(2B)**, 1579 (1996).
[1.19] K. Kim, D. H. Kwon, G. Nallapati, G. S. Lee, *J. Vac. Sci. Technol. A* Pt. 2, **16(3)**, 1509 (1998).
[1.20] S. Hasegawa, T. Tsukaoka, T. Inokuma, Y. Kurata, *J. Non-Cryst. Solids* **240(1-3)**, 154 (1998).
[1.21] S. E. Kim, C. Steinbruchel, *Appl. Phys. Lett.*, **75(13)**, 1902 (1999).
[1.22] H. Yang, G. Lucovsky, *J. Non-Cryst. Solids*, **254**, 128 (1999).
[1.23] N. Yamada, T. Takahashi, *Jpn. J. Appl. Phys., Pt 1*, **39(3A)**, 1070 (2000).
[1.24] K. Numata, T. R. Seha, S.-P. Jeng, T. Tanaka, *Mater. Res. Soc. Symp. Proc.*, **381**, 255 (1995).
[1.25] S.-W. Chung, J.-H. Shin, N.-H. Park, J.-W. Park, *Jpn. J. Appl. Phys. Pt 1*, **38(9A)**, 5214 (1999).
[1.26] P. Sermon, K. Beekman, S. McClatchie, *Vacuum Solutions*, **5**, 31 (1999).
[1.27] V. Rana, R. P. Mandal, M. Naik, D. Yost, D. Cheung, W.F. Yau, *$16^{th}$ Intl. VLSI Mult. Interconn. Conf. (V-MIC)*, September 6-10, Santa Clara, CA (1999).
[1.28] B. C. Auman, *Mat. Res. Symp. Proc.*, **381**, 19 (1995).
[1.29] N. H. Hendricks, K. S. Y. Lau, A. R. Smith, W. B. Wan, *Mat. Res. Symp. Proc.*, **381**, 59 (1995).
[1.30] S. Rogojevic, J. A. Moore, W. N. Gill, *J. Vac. Sci. Technol. A*, **17(1)**, 266 (1999).
[1.31] T. M. Stokich, Jr., W. M. Lee, R. A. Peters, *Mat. Res. Symp. Proc.*, **227**, 103 (1991).
[1.32] A. L. S. Loke, J. T. Wetzel, P. H. Townsend, T. Tanabe, R. N. Vrtis, M. P. Zussman, D. Kumar, C. Ryu, S. S. Wong, *IEEE Trans. Elect. Dev.*, **46(11)**, 2178 (1999).
[1.33] P. H. Townsend, S. J. Martin, J. Godschaix, D. R. Romer, D. W. Smith, Jr., D. Castillo, R. DeVries, G. Buske, N. Rondan, S. Froelicher, J. Marshall, E. O. Shaffer, J.-H. Im, *Mat. Res. Soc. Symp. Proc.*, **476**, 9 (1997).
[1.34] E. E. Marotta, B. Han, *Mater. Res. Soc. Symp. Proc.*, **515**, 215 (1998).
[1.35] J. Pellerin, R. Fox, H.-M. Ho, *Mater. Res. Soc. Symp. Proc.*, **476**, 113 (1997).
[1.36] D. Pingalay, D. Perahia, N. Timble, R. Singh, K. F. Poole, *Proc. Electrochem. Soc.*, **99-7**, 46 (2000).
[1.37] M. DelaRosa, A. Kumar, H. Bakhru, T.-M. Lu, *Mater. Res. Soc. Symp. Proc.*, **565**, 197 (1999).
[1.38] S. Nitta, A. Jain, V. Pisupatti, W. N. Gill, P. C. Wayner, J. L. Plawsky, *Mat. Res. Soc. Symp. Proc.*, **511**, 99 (1998).
[1.39] C. V. Nguyen, K. R. Carter, C. J. Hawker, J. L. Hendrick, R. L. Jaffe, R. D. Miller, J. F. Remenar, H.-W. Rhee, P. M. Rice, M. F. Toney, M. Trollsas, D. Y. Yoon, *Chem. Mater.*, **11(11)**, 3080 (1999).
[1.40] D. Mecerreyes, N. Kamber, E. Huang, V. Lee, T. Magbitang, W. Volksen, C. J. Hawker, R. D. Miller, J. L. Hedrick, *Polym. Prepr.* **41(1)**, 517 (2000).
[1.41] E. M. Zielinski, S. W. Russell, R. S. List, A. M. Wilson, C. Jin, K. J. Newton, J. P. Lu, T. Hurd, W. Y. Hsu, V. Cordasco, M. Gopikanth, V. Korthuis, W. Lee, G. Cerny, N. M. Russell, P. B. Smith, S. O'Brien, R. H. Havemann, *IEDM Tech. Digest*, Dec 7-10, 936 (1997).
[1.42] S.-Q. Wang, in Conf. Proc. VLSI IX, Materials Research Society, 31 (1994).
[1.43] S.P. Murarka, I.V. Verner and R.J. Gutmann, *Copper-Fundamental Mechanics for Microelectronic Applications*, Wiley Interscience, 2000. Chapter 8 and Section 6.3.
[1.44] K.W. Paik and A.L. Ruoff, *Mat. Res. Soc. Symp. Proc.*, **153**, 143 (1989).
[1.45] P.J. Ding, W.A. Lanford, S. Hymes, and S.P. Murarka, *Appl. Phys. Lett.*, 75, 3627 (1994).
[1.46] J.J. Toomey, S. Hymes, and S.P. Murarka, *Appl. Phys. Lett.*, **66**, 2074 (1995).

[1.47] S.W. Hymes, K.S. Kumar, S.P. Murarka, W. Wang, and W.A. Lanford, *Mat. Res. Soc. Symp. Proc.*, **427**, 193 (1996).

## Chapter 2

## LOW-κ INTERLEVEL DIELECTRICS

Considerable research and development is underway to obtain low dielectric constant (low-κ) alternatives to silicon dioxide as an interlevel dielectric (ILD). Materials with low-κ values to be used in the near term, such as polymer materials and organic/inorganic hybrids, target the 2.5 – 3.0 dielectric constant range. Subsequent low-κ technology nodes approach the κ ≤ 2.2 range, implying the incorporation of porosity into one of the currently investigated low-κ dielectrics. Porous materials have many additional challenges for successful processing and integration, requiring extensive experimental evaluation. Such an evaluation examines the effective dielectric constant, water absorption, chemical reactivity, temperature stability, and compatibility of each material with copper metal and barrier materials. More generally, the acceptability of a material, whether dense or nanoporous, is determined by the chemical, mechanical, thermal and electric properties exhibited, as well as integration issues with IC processing requirements.

ILDs can be developed through chemical means to have a wide range of chemical, physical, and electrical properties. In addition, a candidate material in the formation of an IC interconnect must meet rigorous integration requirements. Materials must be able to withstand the physical stresses, chemical environments, and thermal treatments that occur during damascene patterning of interconnect levels. They must exhibit excellent electrical properties, thermal stability, chemical stability, and thermomechanical and thermal stress.

The most important metrics for dielectric materials are electrical properties -- namely low dielectric constant, low dielectric loss, low leakage current, and high breakdown voltage. Many low-κ dielectric materials satisfy these electrical criteria, but have less thermal stability, chemical stability, mechanical strength, and thermal conductivity than $SiO_2$ [2.1]. Tables 2.1 and 2.2. list several of the electrical, chemical, and physical properties that must be balanced for successful integration.

*Table 2.1.* List of electrical and physical properties examined for low-κ films (adapted from [2.2])

| Property | Value |
|---|---|
| Dielectric constant | <3 (preferably <2.5) |
| Dissipation factor at 1 MHz | <0.005 |
| Thermal stability: 1% weight loss in $N_2$ atmos. | >425 °C |
| Moisture absorption | <1% |
| Coefficient of thermal expansion (CTE) | <50 ppm |
| Stress | >±100 MPa |
| Tensile modulus | >1 GPa |
| Tensile strength | >200 MPa |
| Elongation-at-break | >5% |
| Glass transition temperature (Tg) | >400 °C |
| Thermal shrinkage after curing | <2.5% |
| Dielectric breakdown | >1 MV/cm |
| Thermal conductivity | |
| Gas and moisture permeability | |
| Hardness | |
| Density | |

*Table 2.2.* List of properties desired for successful integration of low-κ films (adapted from [2.2])

| Metric | Goal |
|---|---|
| Adhesion (to metal, self-adhesion) | Pass tape test after thermal cycles to 450 °C |
| Gap-fill | No voids at 0.35 µm, aspect ratio = 2 |
| Planarization | >80% (regional) |
| Etch rate | >3 nm/s |
| Step coverage | >80% |
| Reliability of metal, when surrounded by dielectric material | |
| Resistance to solvents, photoresist strippers | |
| Etch selectivity (oxygen plasma resistance) | |
| CMP compatibility | |

Selection of a low-κ material is challenging due to the necessary trade-offs that exist between certain chemical, electrical, and physical characteristics of films. Silicon dioxide is a relatively dense film with an ordered lattice structure that results in good physical stability, thermal conductivity, and electrical performance. Low-κ options are developed to enhance electrical performance by (1) incorporating atoms and bonds that have a lower polarizability, (2) lowering the density of atoms and bonds in the material, or a combination of the two [2.1]. Incorporation of atoms or groups of atoms with lower polarizability such as C, N, F, or molecular groups thereof (see Table 2.3) disrupts the Si-O lattice, resulting in lower dielectric constant, but also lower density, thermal conductivity, physical strength, and thermal stability.

*Table 2.3.* Electronic polarizability of chemical bonds [2.3]

| Bond | Polarizability ($Å^3$) |
|---|---|
| C-C | 0.531 |
| C-F | 0.555 |
| C-O | 0.584 |
| C-H | 0.652 |
| O-H | 0.706 |
| C=O | 1.020 |
| C=C | 1.643 |
| C≡C | 2.036 |
| C≡N | 2.239 |

The effectiveness of a low-κ material to meet the electrical and physical requirements for IC applications can rely heavily on the film deposition methods used (precursors, chemical concentrations) and chosen film structure [2.4]. The following sections describe the advantages and disadvantages of the low-κ options listed in Table 2.4. Each film type has been developed through research aimed at improving one or more material property metrics, while trying to minimize the negative impact on the other metrics.

## 2.1 FLUORINATED GLASSES

Fluorinated silicon dioxide glass (FSG or SiOF) was the first low-κ dielectric implemented in industrial manufacturing. SiOF is typically deposited [2.5-2.6] using plasma-enhanced chemical vapor deposition (PECVD) to obtain films with κ ranging from 3.5-3.7 [2.7]. FSG films can be tailored to contain varied amounts of their Si, O, and F constituents by varying the gaseous precursors used in the deposition reaction chamber.

SiOF may be formed by a variety of methods, each typically combining a fluorine-bearing gas or organic source material, an oxygen plasma, and a silane derivative ($SiR_4$). Several deposition methods are listed in Table 2.5.

Table 2.4. Low-κ dielectric options

| Dielectric Material | Dielectric Constant (κ) | Important Notes |
|---|---|---|
| Silicon Dioxide ($SiO_2$) | 3.9 | IC Processing Standard |
| Fluorinated Oxide (FSG or SiOF) | 3.5-3.7 | Fluorine interaction issues |
| Silsesquioxanes (HSQ, MSQ) | 2.9-3.6 | Liquid spin-on deposition; κ varies with H, C content |
| Carbon-doped Glasses (OSG or SiOC) | 2.7 – 3.0 | CVD or spin-on deposition; κ varies with H, C content |
| Polymers | 2.6-3.2 | Mechanically and thermally less stable than oxides |
| Fluorinated Polymers | 2.1-2.3 | CVD or spin-on deposition; adhesion issues; fluorine interaction issues |
| Porous Media | 1.5-2.0 | Mechanically less stable than non-porous; can be any porous low-κ ($SiO_2$, MSQ, polymer) |

FSG is similar in structure and properties to silicon dioxide. During deposition, fluorine becomes substituted in the $SiO_2$ structure, either at bonding sites due to chemical reaction or interstitial locations due to plasma radical formation/deposition. Fluorine lowers the dielectric constant of $SiO_2$ films by one of several reported mechanisms, including (1) selective replacement of polar Si-OH groups, (2) reduced ionic character of Si-O bonds, and (3) decreased electronic polarizability of the Si-O bond structure [2.9]. The result is a dielectric film that can be "tuned" by the deposition parameters to contain a desired amount of fluorine and the desired film hardness. The dielectric constant dependence upon the ratio of Si-F to Si-O bonds in deposited films is shown in Figure 2.1.

Table 2.5. Reactant for deposition of low-κ SiOF

| Reactant Gases | Plasma Method | Reference |
|---|---|---|
| $Si(OC_2H_5)_4$, $O_2F_6$ | PECVD (Plasma-enhanced CVD) | [2.8] |
| $SiH_4$, $SiF_4$, $O_2$ | RPCVD (Remote plasma CVD) | [2.5],[2.7] |
| $SiH_4$, $N_2O$, $CF_4$ | PECVD | [2.9] |

SiOF films used for intermetal dielectric applications are commonly deposited in the range of 0-10 at% F, producing a dielectric constant of 4.0 (TEOS $SiO_2$) to 3.5. Atomic concentrations are typically quantified by computing the ratio of Si-F to Si-O peaks measured from Fourier Transform Infrared (FTIR) or X-ray Photoelectron Spectroscopy (XPS) spectra [2.9]. F content can also be quantified through the combination of Rutherford Backscattering Spectroscopy (RBS) and nuclear resonance analysis (NRA) [2.10].

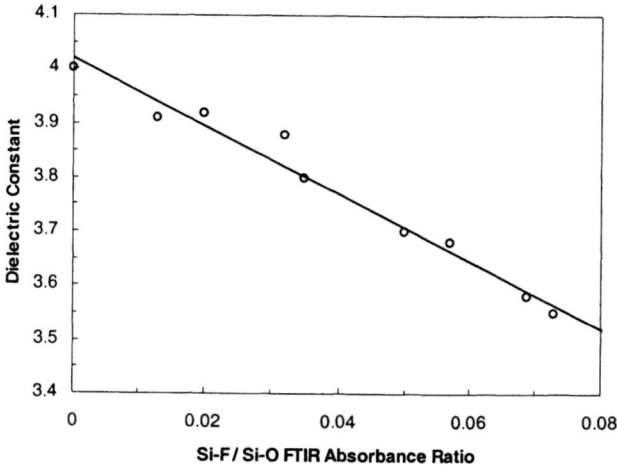

*Figure 2.1.* Dielectric constant vs. the Si-F to SiO FTIR absorbance ratio [2.10]

One concern with SiOF in practical integration is the reactivity of the fluorine within the film. The unbound fluorine can migrate to the dielectric interface and react with conductor or barrier materials [2.11], and the Si-F bond itself may react with water in the SiOF film [2.12]. Addition of large amounts of fluorine to $SiO_2$ films enhances moisture absorption, resulting in an increased κ value. κ is degraded as $H_2O$ reacts with the Si-F bonds in the film, forming Si-OH and HF. Instability of Si-F bonds in $SiO_2$ films is a concern for device reliability [2.13].

## 2.2 SILSESQUIOXANES

Silsesquioxanes are spin-on glasses (SOG) that incorporate the properties of both inorganic ($SiO_2$) and organic (C-based polymer) dielectrics. They have recently found interest as a dielectric option with intermediate values of κ (2.9-3.6) [2.14]. Inorganic/organic hybrid materials use the strong, rigid $SiO_2$ network as a starting point, and lower the κ value by incorporating chemical constituents or voids to lower the film density [2.1]. Silsesquioxanes contain hydrogen and carbon in the form of silicon-terminating bonds (Si-H and/or Si-$CH_3$) and contain small voids due to a preference to form a cage-structure at low reaction temperatures. Figure 2.2 shows the cage structure that is formed at or below 300 °C.

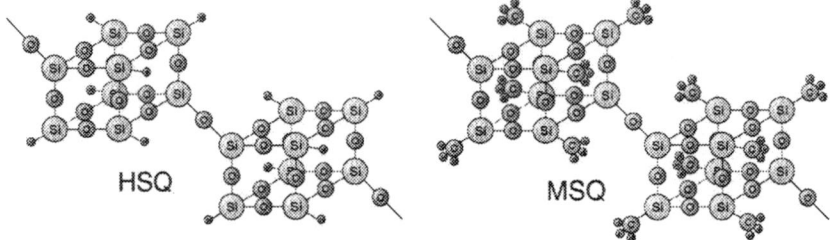

*Figure 2.2* Cage structures of hydrogen silsesquioxane (HSQ) and methyl silsesquioxane (MSQ).

Silsesquioxanes are formed by the liquid-phase polymerization of methyl- and methoxy-substituted silane [$(CH_3)_x(CH_3O)_ySiH_z$, where $x + y + z = 4$] monomers [2.15]. The films are deposited by liquid spin-coating, then heated to initiate the polymerization reaction. Curing polymerizes of the Si-O- backbone of the film, since silsesquioxanes contain no terminal Si-OH or Si-OR groups [2.16]. The reaction causes collapse of the cage structure, forming a networked polymer structure. The extent of amorphous networking depends on the cure time and temperature. Figures 2.3, 2.4, and 2.5 illustrate the observed changes in film thickness, chemistry, and electrical properties with cure temperature.

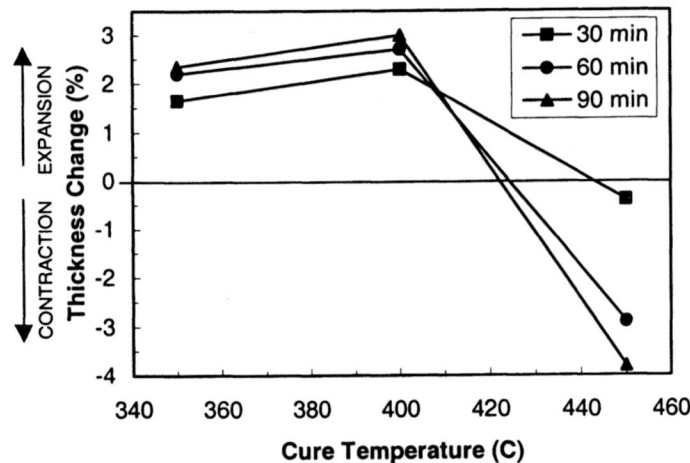

*Figure 2.3.* HSQ film thickness change vs. curing time and temperature [2.16]

The figures show the physical transformation of HSQ during curing, as neighboring Si-O- and Si-H linkages of the cage structures condense to form an amorphous network. FTIR measurements are used to quantify the relative concentration of Si-H or Si-C- bonds in the film and measure hydrogen and carbon loss during amorphization [2.17-2.19]. The cage structure is responsible for the molecular-scale "voids" in HSQ and MSQ that result in reduced film density and dielectric constant. Crosslinking increases the film density, improving the physical strength of the film at the expense of dielectric constant. Consequently, it is important to determine the dielectric constant as a function of the strength of a material, which usually decreases with decreasing density, and then determine which material provides the lowest κ for a given value of its strength.

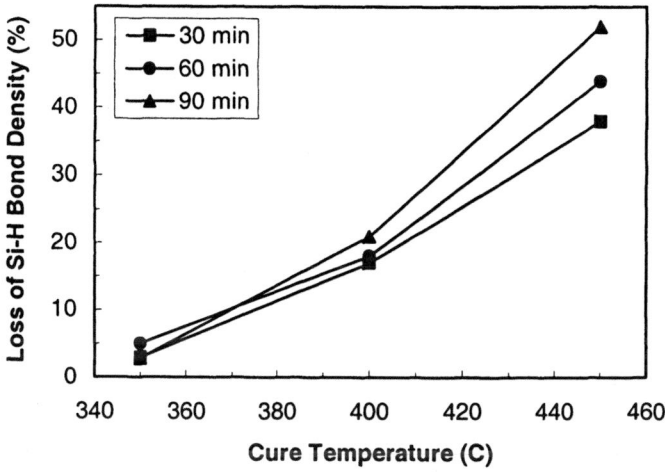

*Figure 2.4.* Loss of Si-H bond density vs. curing time and temperature [2.16]

The tradeoff between thermal stability, physical strength, and electrical properties shown in Figures 2.3-2.5 leads to concern for successful integration of HSQ and MSQ as a near-term low-κ material. Material properties (strength, κ, water content) may vary widely with cure conditions [2.20], and heating or plasma exposure may cause decomposition of methyl or hydrogen from the silsesquioxane cage structure [2.21]. Each of these issues is a challenge for integration into a successful interconnect scheme, where thermal cycles routinely reach temperatures of 450 °C. Recent efforts at improving the integration properties of silsesquioxanes have met with some success. Researchers have successfully improved the thermomechanical stability of silsesquioxanes by moderating the ambient gas used during curing [2.22] and reduced the film reactivity by ion bombardment [2.23].

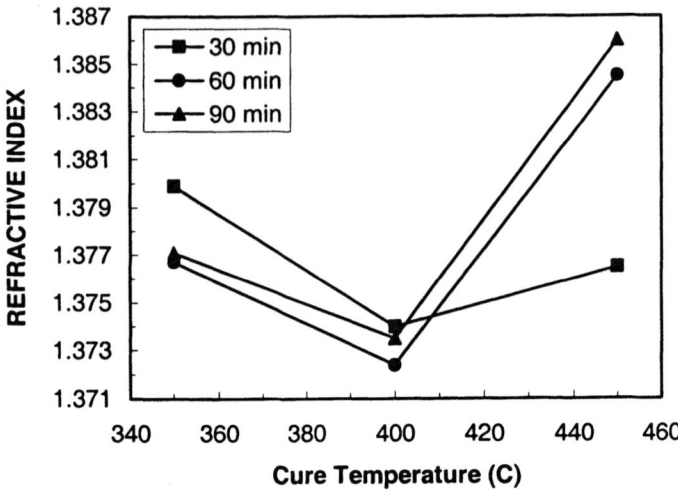

*Figure 2.5.* HSQ film refractive index vs. curing time and temperature [2.16]

## 2.3 ORGANOSILICATE GLASSES

Carbon-doped or "organosilicate" glasses (OSGs) are silicon dioxide films uniformly doped with carbon (SiOC) in an effort to lower dielectric constant. SiOCs are deposited by gas-phase reactions through chemical vapor deposition (CVD) processes, rather than the liquid-phase spin-on/cure approach used with silsesquioxanes. The result is a more controlled polymerization reaction that incorporates carbon into terminal bonds in the $SiO_2$ lattice [2.24]. Carbon can be incorporated in the $SiO_2$ structure during the PECVD process by replacing the silane reactant ($SiH_4$) with a methyl silane derivative [$(CH_3)_xSiH_y$] in the presence of $O_2$ plasma. Methyl silane [$CH_3SiH_3$, known as "1MS"], tri-methyl silane [$(CH_3)_3SiH$, known as "3MS"], and tetra-methyl silane [$(CH_3)_4Si$, known as "4MS"] are three such reactant gases [2.25, 2.26]. During deposition, the methyl groups from the reactant feed become randomly bound to silicon atoms through the formation of C-Si-H radicals [2.27]. Doping a film with carbon terminates some of the silicon bonds within the oxide lattice, leading to a reduction in film density and a reduced number of electrical dipoles present in the film per unit volume [2.28]. The result is lower density carbon-doped oxide film with the structure shown in Figure 2.6 and κ of 2.7-3.0.

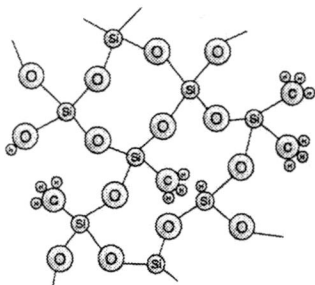

*Figure 2.6:* Amorphous SiOC structure, showing Si bond termination by methyl ($CH_3$) groups or hydrogen

The dielectric constant decreases as the deposition ratio of organic silane to plasma gases is increased (see Figure 2.7). The dielectric constant is tunable as a function of the amount of carbon incorporated into the film and the film density. Both of these parameters are a function of the reactant concentration and the plasma power used during deposition [2.29]. SiOC films are desirable low-κ candidates due to their similarity to oxides, their ability to be deposited by CVD, and the lack of mobile fluorine that may react with interconnect metals [2.29]. OSG also have a high glass transition temperature (~600°C) and low moisture absorption due to their hydrophobic character. As with SiOF and silsesquioxanes, the chemical bonding stability of SiOC can be measured by FTIR, as shown for a 3MS SiOC annealed from 200-600 °C in Figure 2.8.

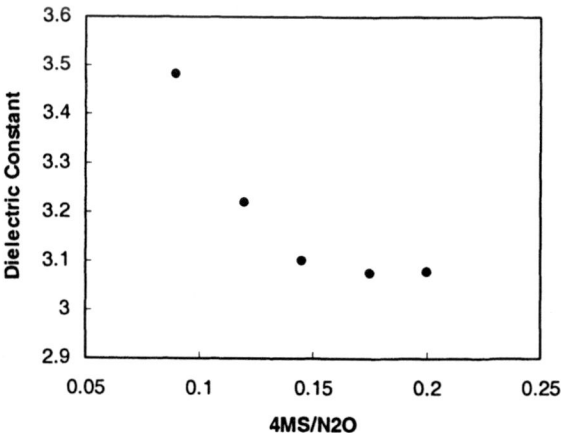

*Figure 2.7.* Variation of dielectric constant with $4MS/N_2O$ ratio for films prepared from polymerization of 4Ms and copolymerization of $4MS/SiH_4$ [2.26]

The films exhibit excellent chemical stability, as the chemical bonding peaks do not decrease in intensity as a result of thermal stress. Stability in the FTIR traces shows a resistance to loss of methyl or hydrogen groups, which would increase the film dielectric constant. Figure 2.9 shows the stability of the SiOC film dielectric constant for the same film under thermal stress conditions.

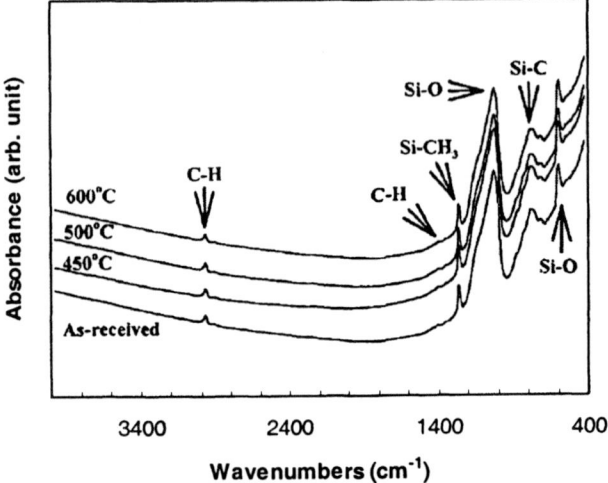

*Figure 2.8.* FTIR absorption spectra of SiOC film annealed at various temperatures for 30 min in an $N_2$ ambient. Adapted from [2.25]

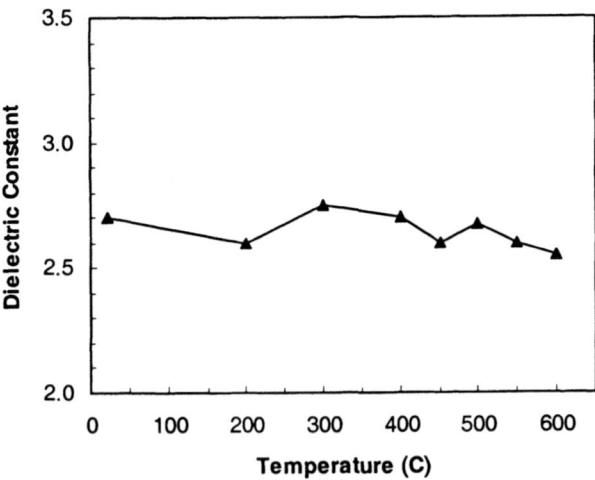

*Figure 2.9.* Dielectric constant vs. annealing temperature for SiOC film. Adapted from [2.25].

SiOC films have been proven to be stable to temperatures of ~650°C at 100% relative humidity [2.30]. Table 2.6 shows that SiOC has the most advantageous combination of thermal and moisture uptake stability of the "doped $SiO_2$" low-κ derivatives.

Table 2.6. Stability of various contents in siloxane-based low-κ dielectrics. Adapted from [2.30].

| Bond | Decomposition temperature (°C) | Moisture resistance |
|---|---|---|
| Si-F (SiOF) | >750 | Poor |
| Si-H (HSQ) | 400 | Poor |
| Si-$CH_3$ (SiOC) | 700 | Good |
| Si-C-Si (SiOC) | >750 | Good |

Overall, SiOC inorganic/organic hybrid materials have received wide acceptance as a reliable near-term low-κ dielectric option with minimal integration difficulties. The main challenge for SiOC integration is that the films contain both inorganic ($SiO_2$-like) and organic (C-like) content, creating challenges in etch and CMP selectivity. Etch chemistries that are designed to remove photoresist will attack the carbon in the SiOC structure, leaving behind silanol (Si-OH) group constituents in the films, leaving a carbon-rich layer on the thin film. Intelligent etch, clean, polish, and patterning strategies must be implemented to protect the physical, chemical, and electrical benefits of the SiOC films.

## 2.4 POLYMERS

The previous discussion of low-κ alternatives has centered around derivatives of $SiO_2$ that use micro-scale voids or lower-polarity ligands to reduce the dielectric constant of $SiO_2$. The use of polymer materials as low-κ ILDs takes a different approach, implementing an organic support that may contain little inorganic material. Polymers have advantageous κ values (2.0-3.0) due to the nature of their bonding and structure. The covalent organic bonding structure of most polymers has less polarization affinity than the Si-O- matrix, reducing the propagation of electrical signal between interconnect lines. Besides the dielectric constant, the most important requirement for polymer ILD candidates is thermal stability, as the polymeric structure is more susceptible to alteration as the temperature is increased.

A wide variety of organic polymer dielectrics can meet the electrical requirements of the κ ~ 2.7 technology node. Polymers can be deposited using either CVD or spin-on polymerisation. CVD has the advantages of a gas-phase reaction with no need for drying or baking of a polymer solvent.

However, few polymers can be successfully deposited by CVD. The deposition method of choice for highly cross-linked, high temperature polymers is spin-coating. ILD candidates for Cu integration require long, cross-linked carbon bonded chains (for strength and thermal stability) and low oxygen content (for low electronic polarizability).

Organic materials can be tailored using synthetic chemistry to be hydrophobic -- to reduce the moisture uptake that is detrimental to their insulating ability. Due to the high temperatures involved in semiconductor processing, successful ILD candidates must also have high temperature stability. The main contributor to the stability of organic dielectrics is the highly stable six-carbon benzene ring. Structures of some organic candidates are shown in Table 2.7 [2.31].

Polyimides [2.36] are linear polymers that contain carbon, oxygen, and nitrogen. The presence of double bonded oxygen or aliphatic nitrogen in the structure degrades performance due to water absorption and higher dielectric constant. Most polyimides do not form extensive cross-linked bonds and are thus labelled "thermoplastics". Thermoplastics, in general, have lower film strength and lower glass transition temperature than densely cross-linked thermosets. In addition, the mobility of the polyimide chains, when coupled with the stresses involved in spin-coating and solvent evaporation, can cause alignment of the polyimide strands, creating anisotropy in out-of-plane physical or electrical properties [2.37].

*Table 2.7.* Structure and properties of several low dielectric constant polymers.

| Polymer | Structure | κ | Glass Transition $T_g$ (°C) | Water Absorption (%) |
|---|---|---|---|---|
| Polyimide [2.2],[2.32] | | 2.7 – 3.1 | 350 | 1.5 – 3.0 |
| Poly(arylene) ether [2.33] | | 2.6 – 2.8 | 260 – 450 | < 0.4 |
| DVS-BCB [2.34] | | 2.65 | > 350 | < 0.2 |
| SiLK [2.4] | | 2.65 | > 450 | < 0.25 |
| Parylene [2.35] | | 2.6 – 2.8 | 420 | < 0.05 |

Polyimides have the advantage that they have well-known synthetic chemistry, and can be tailored using different monomers to have a variety of characteristics such as chain length and oxygen content which affect physical or electrical properties. Fluorine can also be substituted into one of the polyimide block molecules, reducing the film dielectric constant as shown in Figure 2.10.

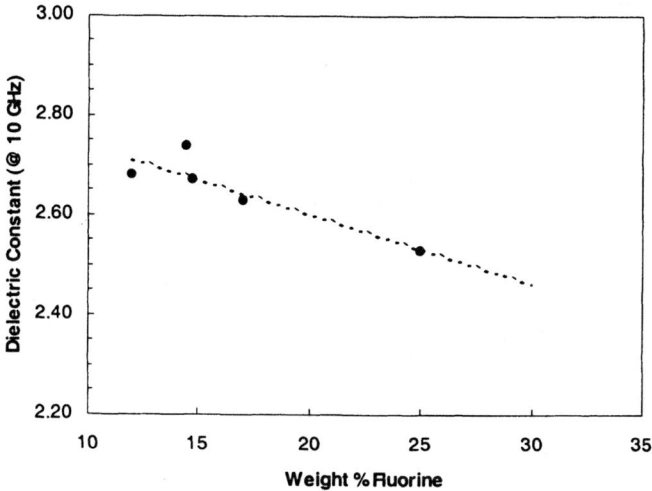

*Figure 2.10*: Polyimide dielectric constant as a function of fluorine content. Adapted from [2.32]

Due to the reactivity of fluorine radicals, weakly-bound fluorine in polyimide materials can cause problems in metal integration. Free fluorine can diffuse through ILD materials and corrode copper. This can happen if the processing temperature approaches the glass transition temperature for the polyimide. The high temperature processing thermal budget is the strongest deterrent for using polyimide low-κ polymers in Cu damascene interconnect processing.

Poly(arylene) ethers are a second group of aromatic polymers. Their name is derived from their *aryl*, or benzene structures, connected by single oxygen atoms, or *ether* linkages. They have high temperature stability due to high content of benzene rings. The oxygen linkage provides flexibility in the chain to prevent thin film cracking and allow maximum crosslinking, but also results in anisotropy in electrical properties. Poly(arylene) ethers have lower dielectric constant and lower water uptake than polyimides due to the absence of carbonyl bonds and much lower oxygen content. A commercial poly(arylene) ether, FLARE, has been developed for the IC market by Allied Signal, Inc. FLARE has been observed to have better thermomechanical stability than HSQ due to less cured film shrinkage [2.38]. Figure 2.11

shows that the enhanced cross-linking of the poly(aryl) ether polymer provides a more rigid and stable film than weaker, more easily decomposed silsesquioxane Si-H bonds.

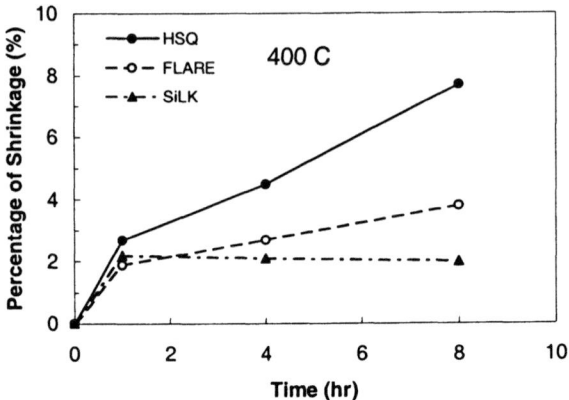

*Figure 2.11*: Annealing time dependence of percentage shrinkage of film thickness for HSQ, FLARE, and SiLK dielectrics annealed at 400 °C [2.38]

DVS-BCB (divinyl siloxane bis-benzocyclobutene) is a hybrid polymer that contains a small amount of silicon in addition to carbon, oxygen, and hydrogen. BCB is a spin on polymer that reacts upon curing to form a highly crosslinked thermoset structure [2.39]. Table 2.7 shows that each BCB monomer has four bonding sites at which to crosslink. This results in good mechanical strength and thermal stability. BCB polymers can also be tailored using a variety of chemical units to achieve desired properties. The inorganic -Si- linkage in the polymer helps provide toughness and increases the glass transition temperature ($T_g$). Table 2.5 also shows that BCB has excellent wafer absorption properties. The dielectric constant of BCB can be adjusted using fluorinated monomers, but the chance for fluorine-induced corrosion, coupled with the sub-400 °C glass transition temperature has not led to significant use of F-BCB as an ILD.

One key advantage of non-fluorinated BCB polymer is that it exhibits excellent barrier properties for hindering copper diffusion under bias conditions [2.40]. Figure 2.12 shows that BCB is second only to PECVD oxynitride at minimizing Cu drift rate through the dielectric. Low electrical permittivity is a key metric for integration when selecting the necessary barrier or cap layers for a chosen dielectric.

# LOW-K INTERLEVEL DIELECTRICS

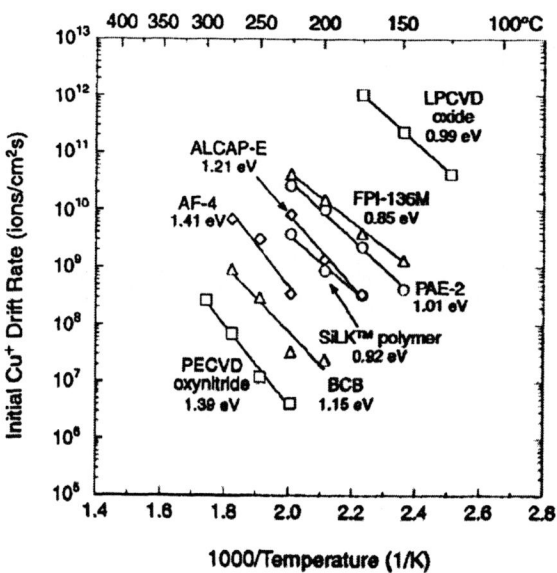

*Figure 2.12*: Arrhenius plot of initial Cu+ drift rates in various dielectrics [2.40].

Another spin-on aromatic hydrocarbon polymer developed for IC applications is SiLK [2.34, 2.41]. SiLK consists almost completely of aromatic linkages, as shown in Table 2.8. The SiLK aromatic bonds provide the highest temperature stability possible for an organic material and makes the structure very inert to chemical attack. A lack of oxygen or nitrogen makes SiLK hydrophobic, resulting in low water uptake. SiLK, like BCB, is prepared from a partially polymerized solution by spin-coating and curing monomer solutions to form a densely crosslinked film. When cured at 450 °C, the dielectric constant of SiLK is 2.65.

SiLK polymer has the most desirable balance of physical, thermal, and chemical properties of the polymers discussed [2.42]. The organic structure of SiLK results in low physical strength, hardness, modulus, and toughness. However, the film is quite stable at high temperatures, as indicated by film shrinkage at 400°C shown in Figure 2.11. For these reasons, SiLK has found industry acceptance for initial integration into multilevel interconnects as depicted in Figure 2.13.

The main integration concerns regarding the low-κ polymers mentioned to this point (polyimides, poly(arylene) ethers, BCB, and SiLK) are their lower temperature stability, thermal conductivity, and mechanical stability compared to $SiO_2$. Currently, the controlled collapse chip connection (C4) technique used in chip packaging requires the interconnect to withstand

*Figure 2.13.* SEM cross section of IBM interconnect structure composed of five layers of copper wires in SiLK dielectric with an additional top two layers of copper in silicon dioxide (SEM courtesy of IBM) [2.43]

*Table 2.8:* Summary of SiLK dielectric properties [2.4]

| Property | Value |
|---|---|
| Dielectric constant | 2.65 |
| Voltage breakdown | 4 MV/cm |
| Leakage current at 1 MV/cm | 0.33 nA/cm$^2$ |
| Refractive index at 632.8 nm | 1.63 |
| Moisture uptake at 20°C, 80% RH | 0.24% |
| Thermal stability | >425 °C |
| Weight loss at 50°C | 0.7 wt% hr |
| Thermal conductivity at 25, 125 °C | 0.19, 0.23 W/mK |
| Glass transition temperature | >490 °C |
| Young's modulus | 2.45 GPa |
| Strength | 90 MPa |
| Ultimate strain | 11.5% |
| Hardness | 0.38 GPa |
| Toughness | 0.62 MPa m$^{1/2}$ |
| Residual stress at RT | 56 MPa |
| Coefficient of thermal expansion | 66 ppm/°C |

temperatures above 400°C [2.44], which is above the glass transition temperature ($T_g$), the temperature at which the material changes from a solid to a supercooled liquid, for several of the polymers listed in Table 2.5. Polymer films do not conduct heat away from the interconnect as well as $SiO_2$, and are also mechanically weaker due to their amorphous, carbon-based structure. However, polymers are finding a great deal of interest in the industry due to their low dielectric constant, ease of deposition, and ability to reduce interconnect stress. Challenges must be met in order to successfully

# LOW-K INTERLEVEL DIELECTRICS

integrate a polymer or other low-κ material in an interconnect scheme. The chemical demands of dry and wet etching, the thermal demands of metal anneals, and the physical demands of CMP are stringent specifications on these materials, and may require creative approaches to integration such as capping layers or processing hardmasks. In Section 5.4, these requirements are highlighted for a specific low-κ OSG.

The polymers Parylene-N and Parylene-F are vapor deposited polymers with sufficiently low dielectric constants to be considered as ILDs. Parylene-N has a dielectric constant of about 2.65 and Parylene-F (often referred to as AF-4) is reported to be as low as 2.25 [2.45]. The structures of Parylene-N and Parylene-F are shown in Figure 2.14.

In Parylene-F fluorine replaces all four of the hydrogen atoms in the aliphatic groups of the repeat unit. Apparently, the presence of fluorine in AF-4 accounts for its lower dielectric constant. However, the fluorine also may be a problem in IC interconnect applications, where fluorine can interact with copper, aluminium and liners, particularly in the presence of trace levels of moisture. Diffusion barriers are needed to minimize fluorine diffusion if fluorinated low-κ ILDs are employed [2.46].

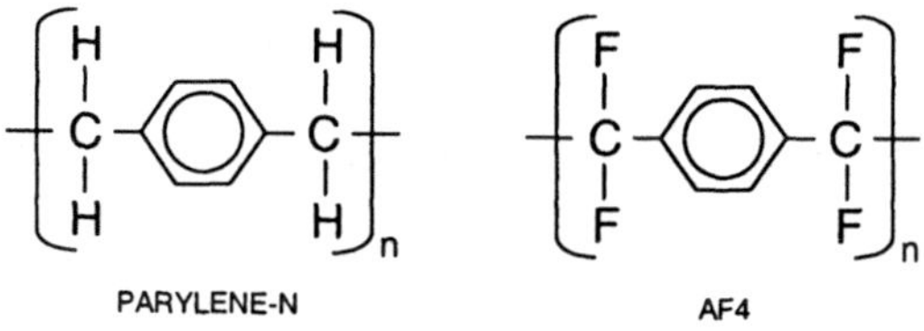

*Figure 2.14*: Chemical Structure of various parylenes [2.45].

The deposition rate of vapor deposited polymers such as parylene is often too low (<100 nm/min) to be suitable for ILDs in single-wafer systems. However, the thermal stability and growth rate of Parylene has been shown to increase significantly by low temperature deposition [2.47]. Models of Parylene deposition rates have shown that using the proper mechanism of adsorption on the surface of the film is a critical factor in obtaining good agreement with experimental results [2.48]. Multilayer adsorption [2.49] simulated experimental data well over a broad range of temperature.

Parylene-F (or parylene AF-4) has a lower dielectric constant (κ = 2.25) and better thermal stability than Parylene-N, which is attractive for ILD applications. Double-level metal VLSI structures have been fabricated using AF-4 with a significant reduction of line-to-line capacitance and acceptable electromigration lifetimes and leakage currents [2.45]. The thermal stability of Parylene-F is markedly better than Parylene-N, decomposing at 530°C in nitrogen and at 400°C in air [2.54], [2.55]. The reported growth rate for

Parylene-F at low temperatures is 180 nm/min at -13°C -- with the rate increasing with decreasing temperature [2.56]. These results are consistent with the predictions and observations described earlier for Parylene-N [2.47].

Integration issues with copper damascene patterning have been considered by determining the plasma etch rates [2.50] and mechanical properties [2.51] of Parylene-N. Annealing can increase the hardness of Parylene-N by a factor of three [2.52]. Other researchers also have studied the various aspects of integrating Parylene-F in integrated circuits [2.57]-[2.61]. However, Novellus [2.62] has recently decided, after several years of effort, that Parylenes require an integration sequence more expensive than other low-κ dielectrics such as carbon doped glass and polymers like SiLK. In spite of some attractive material properties, use of parylene for copper/low-κ interconnect structures for widespread manufacturing applications has not been pursued actively.

## 2.5 FLUORINATED HYDROCARBONS

Labelle, Lau and Gleason [2.63] have described pulsed plasma enhanced experiments using $CH_2$, $F_2$, $C_2$, $H_2F_4$ and $CHClF_2$ for precursors in a search for lower-κ polymers (κ<2.4). Fluorocarbon films derived from hexafluoro propylene oxide (HFPO) have dielectric constants of 2.0 to 2.4. A pulsed, rather than continuous, plasma system enables a lower ratio of ions to reactive neutrals during the plasma off period than during the plasma on period. Therefore the process equilibrium shifts to favor film deposition from reactive neutrals, causing deposition of a low-density, highly cross-linked low-κ film. Labelle et al. also note that one can improve the films and decrease the density of unpaired electrons (dangling bonds) with increased pulse off time. Film composition, thermal stability and electrical properties for these flurocarbon films are the key metrics affecting mechanical, thermal, and electrical properties.

The lowest known dielectric constant of a dense material is 1.9, which reported for polytetrafluoroethylene, Teflon. The chemical formulas for the monomer of Teflon and Teflon – AF are shown in Figure 2.15. Teflon - AF seems to have better properties for microelectronic applications than Teflon in that the additional group in its structure inhibits crystallization and it appears to make it free of creep. However, Teflon dissociates at 400°C in nitrogen, releasing fluorine which can be deleterious in microelectronic applications.

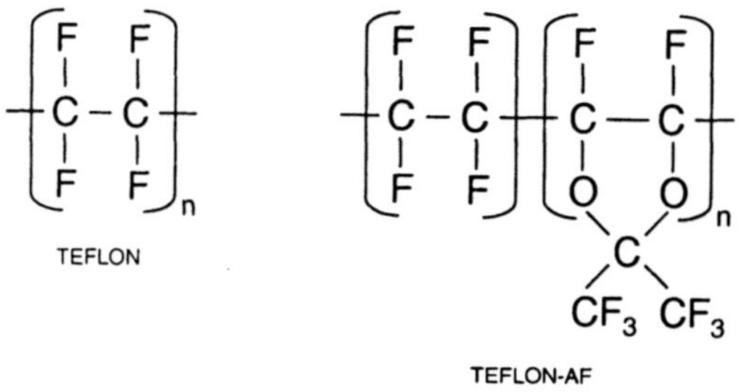

Figure 2.15. Chemical formulas for the monomers of Teflon and Teflon–AF.

The very desirable value of $\kappa=1.9$ may be offset by the integration problems encountered with Teflon. These problems involve thermal stability, mechanical stability and adhesion. Interactions with other materials have been reported. They indicated that no chemical interaction occurs between Ag, Au Cu and Teflon-AF, but a strong interaction was observed with Al [2.64],[2.65]. They showed that adhesion can be improved by treating the interface with reactive ion assisted interface bonding with subsequent annealing.

## 2.6 NANOPOROUS SILICA FILMS

Nanoporous low dielectric constant films are very attractive for ILD applications to reduce $\kappa$ to a minimum, with various materials having been processed in a nanoporous form to reduce $\kappa$. While increasing porosity reduces the dielectric constant, the mechanical, thermal and chemical properties are often adversely affected. Thus the crucial questions are (1) what are the mechanical, thermal and chemical properties of a material for a given value of the dielectric constant, and (2) how are these properties affected by the way the material is processed.

Nanoporous silica films (xerogels) have been investigated more extensively than most other materials, since nanoporous silica is essentially the same material as the thermally grown and PECVD $SiO_2$ that has been used in integrated circuits for many years. The dense material has good thermal stability, high thermal conductivity, modulus, hardness, electrical properties and well known etching characteristics. Beyond its potential for use as a low-$\kappa$ film in integrated circuits with metal interconnects, the low and controllable refractive index, negligible absorption and scattering, and controllable thickness make xerogels an excellent cladding material for optical waveguides. Optical waveguides have been proposed as an alternative to metal interconnects in the future, as discussed in Chapter 8.

Xerogel films for mechanical, chemical, electrical testing or for waveguide development can be made by a multistep procedure [2.73]-[2.75]. The porosity can be controlled by regulating the rate of evaporation of solvent (ethanol) during spin coating or by using a mixture of TEOS and ethylene glycol in the preparation process. Both the TEOS/ethylene glycol ratio, and sintering after the films are made seem to have a profound effect on the microstructure and the properties of the material. Surface modification, during which methyl groups are added on the surface, renders the films hydrophobic even under severe environmental conditions [2.76].

The use of porous materials leads to issues regarding the effect of porosity on their adhesion to contiguous layers, their mechanical stability, their ability to transfer heat and their chemical and electrical interaction with adjacent layers. All of these aspects of the use of porous materials depend not only on porosity, but also on processing conditions which lead to different microstructures. The connectivity of the porous structure, i.e.: open or closed cell pores, is believed to have a significant effect on how the films behave, especially with respect to chemical interactions with neighboring layers. Since closed pore structures are limited to porosity of 30% or less, it is important to determine if chemical interactions can be inhibited in open pores materials such as one has with nanoporous silica at porosities above 30%, and with correspondingly lower dielectric constants. Determination of the relationship between mechanical, thermal, electrical and chemical interaction properties and dielectric constant is highly desirable.

A major concern with dielectric films is how susceptible they are to interaction with contiguous layers such as copper and tantalum. Tantalum is a potential diffusion barrier in copper/low-κ integrated circuits. Upon annealing in nitrogen [2.77], diffusion of tantalum into xerogel is not observed when the tantalum is prevented from oxidizing using a silicon nitride cap layer. However, the adhesion of tantalum to nanoporous xerogel is reduced compared to $SiO_2$ due to reduced contact area and the chemistry of the interface between the materials [2.77]. Improved adhesion, perhaps using an adhesion promoter, is necessary if tantalum is to be used as a diffusion barrier for copper.

The potential for high porosity with correspondingly high internal surface area also causes concern that diffusion of metals, in particular copper, could be much faster than through a solid material such as thermally grown $SiO_2$. Surface diffusion may be more rapid, and dominate over bulk diffusion at low temperatures in highly porous substrates. Therefore, diffusion of copper through xerogels has been examined by investigating the susceptibility of copper penetration under bias-temperature-stress (BTS) both with and without an electric field present [2.78],[2.79]. Thermal diffusion, without an electric field, does not occur after annealing for 2 hrs. at 650°C, based on RBS analysis. Another sample annealed at 450°C revealed no detectable diffusion with secondary ion mass spectrometry (SIMS), which is an order of magnitude more sensitive than RBS.

Since copper is known to be a fast diffusing species in thermally grown $SiO_2$ when subjected to an electric field (i.e., field-enhanced diffusion), annealing also was done under these BTS conditions. According to Loke et al. [2.80], the leakage current through capacitors fabricated with copper is a good indication of the magnitude of copper diffusion. Data of this kind have been reported to study copper diffusion in xerogels with BTS. Contrary to expectations it was found that the leakage current through xerogels is an order of magnitude less with xerogels in Cu/Xerogel samples than in Al/xerogel experiments.

Capacitance-voltage (C-V) testing also is used to detect charges in a dielectric, including those from the diffusion of metal ions. C-V testing of BTS samples shows that xerogel films have charge traps (or there is a significant amount of charge injection) which precludes detecting copper ion diffusion by this method [2.77-2.79]. It is possible that the inhibition of copper diffusion in xerogels even under the influence of an electric field is caused by the non-wetting nature of copper on xerogel, since the highly branched silica network creates high tortuosity for solid-phase diffusion and inactive chemistry at the xerogel surface.

The main challenges to practical integration of porous xerogel films remain hardness and water adsorption. For a given dielectric constant either sintered or templated, silica has a higher modulus than films made of any other material including polymers or hybrids [2.82] as shown in Figure 2.16.

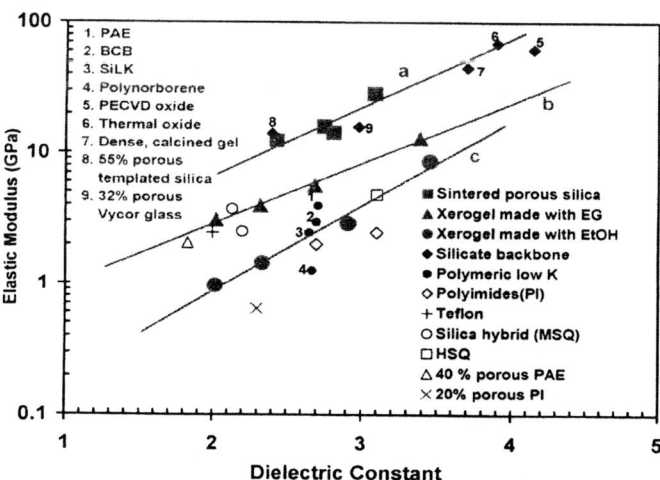

*Figure 2.16:* Relationship of elastic modulus to dielectric constant for different low-κ dielectrics

Small amounts of water, which has a dielectric constant of 72, can render low-κ porous films useless by raising their dielectric constants significantly [2.76, 2.87]. A quartz crystal microbalance (QCM), which is one of the most sensitive methods for measuring small changes of mass, has been used to determine quantitatively the amount of water absorbed up by nanoporous silica as a function of porosity and surface modifier under conditions of 100% relative humidity [2.76]. Silylation (surface modification) is very effective in reducing moisture content and making nonoporous silica films hydrophobic by replacing OH groups by non-polar ones [2.76, 2.85, 2.86].

## 2.7 OTHER POROUS MATERIALS

Several research and development efforts have been made to introduce air into existing dielectric materials to produce porous materials. As described in the previous section, porous silica [2.66] has been demonstrated to have a tunable κ (depending on the porosity) of between 1.5 and 2.5. Porous inorganic and organic materials have also been made using two-phase systems, where one phase is selectively volatilized following deposition, leaving the structure porous. This method, called "templating" has been used to generate nanoporous methyl silsesquioxanes with κ from 1.9 to 2.2 [2.67-2.68],[2.72]. The general approach used to make closed-pore nanoporous MSSQ was to spin-cast a low molecular weight solution of MSSQ and a polycaprolactone (PCL) polymer onto a silicon wafer, thermally cure at temperatures as high as 250 °C to form a highly cross-linked organic/inorganic polymer hybrid, and then decompose the organic polymer at temperatures between 250 °C to 430 °C, the fragments of which diffuse out of the matrix, leaving a nanoporous foamed structure. Researchers have used this approach to obtain films with porosities up to but not exceeding 30%, but quantitative values of thermal and mechanical properties were not reported.

Porosity can also be incorporated into low-κ polymers to further reduce the dielectric constant. Non-porous polyimide films are anisotropic with respect to both the coefficient of thermal expansion and mechanical properties as indicated by Young's modulus [2.70]. This anisotropy is attributed to the bonding and microstructure of the films. However polyimides can be foamed [2.71], which creates a nanoporous, closed pore structure with up to 30% porosity. The introduction of porosity reduces the anisotropy of the in-plane and out-of-plane dielectric constants, and that dielectric constants close to 2.0 could be obtained with polyimide films. Thermal treatment below the glass transition temperature, $T_g$, results in a foam type structure with closed pores up to 30% porosity, while such treatment above $T_g$ leads to a collapsed structure. While data was not given on the thermal conductivity or modulus of these films, one would expect that they decreased significantly as the porosity increased and dielectric constant decreased [2.73].

Current efforts are focused on developing porosity in a densely crosslinked thermoset polymer such as SiLK. Early demonstrations of porous SiLK have κ values that range from 1.5 – 2.0, depending on porosity. In general, porous options have low dielectric constant but also very low fracture toughness and adhesion strength. The thermal conductivity of dielectric films is also important as it determines their ability to transfer heat by conduction. This property depends strongly on: (1) the degree of porosity, (2) the chemical composition of the film (organic or inorganic), and (3) the film microstructure. One of the attractive features of thermally grown $SiO_2$, which has contributed significantly to its use over a long period of time, is that its thermal conductivity is significantly higher than other potential dielectric films. Thus, it is not surprising that for a given value of the dielectric constant, nanoporous silica has a higher thermal conductivity for a given dielectric constant than any other dense or nanoporous dielectric film, as shown experimentally [2.84].

Figure 2.17 gives a comparison between sintered xerogels processed by single solvent preparation and other nanoporous films. For reference, the calculated maximum value to an $SiO_2$-based porous material is shown. With the exception of Teflon, the thermal conductivity of the polymers are lower than that of sintered Xerogels for the same value of dielectric constant. Figure 2.18 shows that sintered porous silica compares favorably with both xerogels using an ethanol process and polymer films. BCB, SiLK and polyimide are particularly highlighted.

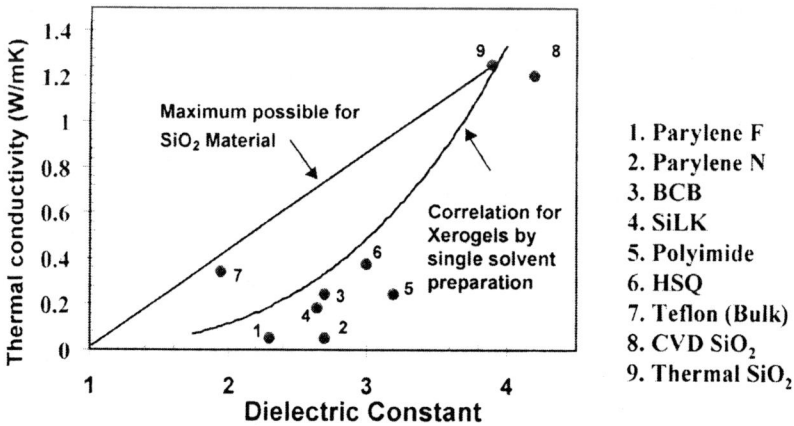

*Figure 2.17:* Thermal conductivity of various low-κ dielectrics compared to xerogels and sintered nanoporous $SiO_2$

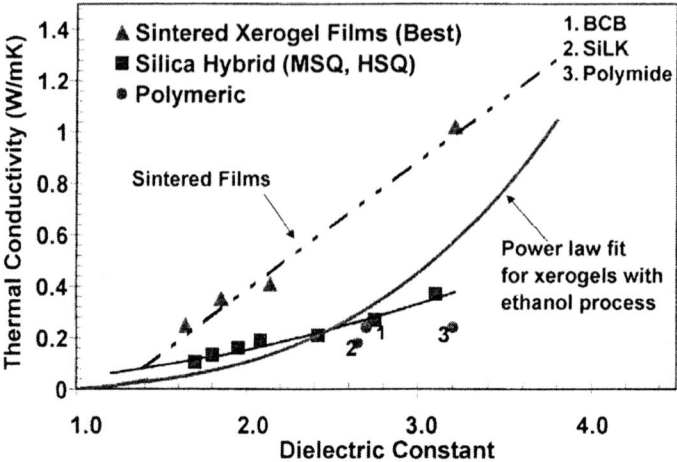

*Figure 2.18:* Thermal conductivity of sintered porous silica and selected polymers as a function of dielectric constant

Porosity leads to challenges in fabrication and integration that must be addressed before these materials can be seriously considered manufacturable [2.69]. These challenges have resulted in ultra low-κ dielectrics being projected later in the 2001 ITRS compared to the 1999 ITRS.

## 2.8 REFERENCES

[2.1] M. Morgen, E.T. Ryan, J.-H Zhao, C. Hu, T. Cho, and P.S. Ho, *Annu. Rev. Mater. Sci.*, **30**, 645 (2000).

[2.2] G. Maier, *Prog. Polym. Sci.*, **26**, 3 (2001).

[2.3] K.J. Miller, H.B. Hollinger, J. Grebowicz, and, B. Wunderlich, *Macromolecules*, **23**, 3855 (1990).

[2.4] S.J. Martin, J.P. Godschalx, M.E. Mills, E.O. Shaffer II, and P.H. Townsend, *Adv. Mater.*, **12(23)**, 1769, (2000).

[2.5] S.-M. Lee, M. Park, K.-C. Park, J.-T. Bark, and J. Jang, *Jpn. J. Appl. Phys., Part I*, **35(2B)**, 1579 (1996).

[2.6] K. Kim, D.H. Kwon, G. Nallapati, and G.S. Lee, *J. Vac. Sci. Technol. A*, **16(3)**, Pt. 2, 1509 (1998).

[2.7] S. Hasegawa, T. Tsukaoka, T. Inokuma, and Y. Kurata, *J. Non-Cryst. Solids*, **240(1-3)**, 154 (1998).

[2.8] T. Usami, K. Shimokawa, M. Yoshimura, *Jpn. J. Appl. Phys.*, **33(1B)**, 408 (1994).

[2.9] F. Iacona, G. Casella, F. La Via, S. Lombardo, V. Raineri, and G. Spoto, *Microelectronic Engineering*, **50**, 67 (2000).

[2.10] D.R. Denison, J.C. Barbour, and J.H. Burkart, *J. Vac,. Sci. Technol. A*, **14(3)**, 1124 (1996).

[2.11] S.E. Kim and C. Steinbruchel, *Appl. Phys. Lett.* **75(13)**, 1902 (1999).

[2.12] H. Yang and G. Lucovsky, *J. Non-Cryst. Solids*, **254**, 128 (1999).

[2.13] N. Lifshitz and G. Smolinsky, *IEEE Electron Device Lett.*, **12**, 140 (1991).
[2.14] K. Numata, T.R. Seha, S.-P. Jeng, T. Tanaka, *Mater. Res. Soc. Symp. Proc.*, **381**, 255 (1995).
[2.15] N. Yamada and T. Takahashi, *Jpn. J. Appl. Phys., Pt 1*, **39(3A)**, 1070 (2000).
[2.16] Y.K. Siew, G. Sarkar, X. Hu, J. Hui, A. See, and C.T. Chua, *Journal of The Electrochemical Society*, **147(1)**, 335 (2000).
[2.17] A. Provatas and J.G. Matisons, *TRIP*, **5(10)**, 327 (1997).
[2.18] W.-J Li, C.-J Yang, and W.-C. Chen, *Mat. Res. Soc. Sym[p] Proc.*, **612**, D5.1.1 (2000).
[2.19] C.Y. Wang, Z. X. Shen, J.Z. Zheng, *Applied Spectroscopy*, **54(2)**, 209 (2000).
[2.20] J.-K. Lee, K. Char, H.-J. Kim, H.-W. Rhee, H.-W. Ro, D.Y. Yoo, and D.Y. Yoon, *Mat. Res. Soc. Symp. Proc.*, **512**, D3.6.1 (2000).
[2.21] S.-W. Chung, J.-H. Shin, and N.-H. Park, *Jpn. J. Appl. Phys. Pt 1*, **38(9A)**, 5214 (1999).
[2.22] H.-C. Liou, E. Dehate, J. Duel, and F. Dall, *Mat. Res. Soc. Symp. Proc.*, **612**, D5.12.1 (2000).
[2.23] T.C. Chang, M.F. Chou, Y. J. Mei, J.S. Tsang, F.M. Pan, W.F. Wu, M.S. Tsai, C. Y. Chang, F.Y., Shih, and H.D. Huang, *Thin Solid Films*, **332**, 351 (1998).
[2.24] H. Yang, D.J. Tweet, L. H. Tecker, W. Pan, D. R. Evans, and S.-T Hsu, *Mat. Res. Soc. Symp. Proc.*, **612**, D3.3.1 (2000).
[2.25] Z.-C. Wu, A.-W. Shjiung, C.-C. Chiang, W.-H. Wu, M.-C. Chen, S.-M. Jang, W. Chang, P.-F. Chou, S.-M. Jang, C.-H. Yu, and M.-S. Liang, *Journal of The Electrochemical Society*, **148(6)**, F127 (2001).
[2.26] L.M. Han, J.-S. Pan, S.-M. Chen, N. Balasubramanian, J. Shi, L.S. Wong, and P.D. Foo, *Journal of The Electrochemical Society*, **148(7)**, F148 (2001).
[2.27] M.J. Loboda, *Microelectronic Engineering*, **50**, 15 (2000).
[2.28] P. Sermon, K. Beekman, and S. McClatchie, *Vacuum Solutions*, **5**, 31 (1999).
[2.29] V. Rana, R.P. Mandal, M. Naik, D. Yot, D. Cheung, and W.F. Yau, *$16^{th}$ Intl. VLSI Mult. Interconn. Conf. (V-MIC)*, September 6-10, Santa Clara, CA (1999).
[2.30] T. Furusawa, D. Ryuzaki, R. Yoneyama, Y. Homma, and K. Hinode, *Journal of The Electrochemical Society*, **148(9)**, F175 (2001).
[2.31] L. Peters, *Semicond. Intl.*, **21(11)**, 64 (1998).
[2.32] J.O. Simpson and A.K. St. Clair, *Thin Solid Films*, **308-309**, 480 (1997).
[2.33] N.H. Hendricks, K.S.Y. Lau, A.R. Smith, and W.B. Wan, *Mat. Res. Symp. Proc.*, **381**, 59 (1995).
[2.34] T.M. Stokich, Jr., W.M. Lee, and R.A. Peters, *Mat. Res. Symp. Proc.*, **227**, 103 (1991).
[2.35] S. Rogojevic, J.A. Morre, and W.N. Gill, *J. Vac. Sci. Technol. A*, **17(1)**, 266 (1999).
[2.36] B.C. Auman, *Mat. Res. Symp. Proc.*, 381, 19 (1995).
[2.37] H.-C. Liou, P.S. Ho, and B. Tung, *Journal of Applied Polymer Science*, **70**, 273 (1998).
[2.38] Z.-C. Wu, X.-W. Shiung, R.-G. Wu, Y.-L. Liu, W.-H. Wu, B.-Y. Tsui, M.-C. Chen, W. Chang, P.-F. Chou, S.-M. Jang, C.-H. Yu, and J.-S. Liang, *Journal of The Electrochemical Society*, **148(6)**, F109 (2001).
[2.39] M.E. Mills, P. Townsend, D. Castillo, S. Martin, and A. Achen, *Microelectronic Engineering*, **33**, 327 (1997).
[2.40] A.L.S. Loke, J.T. Wetzel, P.H. Townsend, T. Tanabe, R.N. Vrtis, M. P. Zussman, D. Kumar, C. Ryu, and S.S. Wong, *IEEE Trans. Elect. Dev.*, **46(11)**, 2178 (1999).
[2.41] P.H. Townsend, S.J. Martin, J. Godschaix, D.R. Romer, D. W. Smith, Jr., D. Cstillo, R. DeVries, G. Buske, N. Rondan, S. Forelicher, J. Marshall, E.O. Shaffer, and J.-H. Im, *Mat. Res. Soc. Symp. Proc.*, **476**, 9 (9197).
[2.42] J.C. Maisonobe, G. Passemad, C. Lacour, Lecornec, P. Motte, P. Noel, and J. Torres, *Microelectronic Engineering*, **50**, 25 (2000).

[2.43] R.D. Goldblatt, B. Agarwala, M.B. Anand, E.P. Barth, G.A. Biery, Z.G. Chen, S. Cohen, J.B. Connolly, A. Cowley, T. Dalton, S.K. Das, C.R. Davis, A. Deutsch, C. DeWan, D.C. Edelstein, P.A. Emmi, C.G. Faltermeier, J.A. Fitzsimmons, J. Hedrick, J.E. Heindrenreich, C.K. Hu, J.P. Hummel, P. Jones, E. Kaltalioglu, B.E. Kiastenmeier, M. Chrisnan, W.F. Landers, E. Liniger, J. Liu, N.E. Lutig, S. Malhotra, D.K. Manger, V. McGahay, R. Mih, H.A. Nye, S. Purushothaman, H.A. Rathore, S.C. Seo, T.M. Shaw, A. H. Simon, T.A. Spooner, M Stetter, R.A. Wachnik, and J.G. Ryan, *Proc. Int., Interconnect Tech. Conf.*, San Francisco, CA, June 5-7, 2612 (2000).
[2.44] E.E. Marotta and B. Han, *Mater. Res. Soc. Symp. Proc.*, **515**, 215 (1998).
[2.45] J.F. Gaynor and A.R.K. Ralston, Semiconductor International, 73 (Dec. 1997).
[2.46] M. DelaRosa, A. Kumar, H. Bakhru and T.-M. Lu, Mat. Res. Soc. Proc., **564**, 559 (1999).
[2.47] S. Ganguli, H. Agrawal, B. Wang, J.F. McDonald, T.-M. Lu, G.-R Yang, and W.N. Gill, *J. Vac. Sci. Technol.* A **15(6)**, 3138, 1997.
[2.48] S. Rogojevic, J.A. Moore, and W.N. Gill, *J. Vac. Sci. Technol.* A **17(1)**, 266, 1999.
[2.49] Brunauer, Emmett and Teller, *Journal of Amer. Chem. Soc.*, **60**, 309, 1938.
[2.50] R.D. Tacito and C. Steinbruchel, *J. Electrochem. Soc.*, **143(6)**, 1974, 1996.
[2.51] C. Chiang, A.S. Mack, C. Pan, Y.-L Ling, and D.B. Fraser, *Mat. Res. Soc. Symp. Proc.*. **381**, 123, 1995.
[2.52] R.J. Gutmann, T.P. Chow, D.J. Duquette, T.-M. Lu, J.F. McDonald and S.P. Murarka, *Mat. Res. Soc. Symp. Proc.*, **381**, 177, 1995.
[2.53] K.J. Taylor, S.-P. Jeng, M. Eissa, J. Gaynor, and H. Nguyen, in Materials for Advanced Metallization, MAM '97, *Proceedings of the Second European Workshop on Materials for Advanced Metallization, Villard de Lans*, France, March 16-19, 1997, 59, 1997.
[2.54] B.L. Joesten, *Journal of Applied Polymer Science*, **18**, 439, 1974.
[2.55] K.R. Williams, *Journal of Thermal Analysis*, **49**, 589, 1997.
[2.56] A.S. Harrus, M.A. Plano, D. Kumar, and J. Kelly, *Mat. Res. Soc. Symp. Proc.*, **443**, 21, 1997.
[2.57] A.R.K. Ralston, J.F. Gaynor, A. Singh, L.V. Le, R.H. Havemann, M.A. Plano, T.J. Cleary, J. Wing, and J. Kelly, *Jpn. Soc. Appl. Phys.*, **81**, 1997.
[2.58] J. Gaynor, J. Chen, H. Nguyen, G. Brown, K. Taylor, J.D. Luttmer, M.A. Plano, T. Cleary, J. Wing, and J. Kelly, *The Electrochemical Society Proceedings*, **97-98**, 157, 1997.
[2.59] R. Sutcliffe, W.W. Lee, J.F. Gaynor, J.D. Luttmer, D. Martini, J. Kelber, and M.A. Plano, *Applied Surface Science*, **126**, 43, 1998.
[2.60] H. Bakhru, A. Kumar, T. Kaplan, M. Delarosa, J. Fortin, G.-R. Yang, T.-M. Lu, S. Kim, C. Steinbruchel, X. Tang, J.A. Moore, B. Wang, J. McDonald, S. Nitta, V. Pisupatti, A. Jain, P. Wayner, J. Plawsky, W.N. Gill, and C. Jin, *Low Dielectric Constant Materials IV, Mat. Res. Soc. Symp. Proc.*, **511**, 125, 1998.
[2.61] G.-R. Yang, Y.-P. Zhao, B. Wang, E. Barnat, J. McDonald, and T.-M Lu, *Applied Physics Letters*, **72(15)**, 1846, 1998.
[2.62] Technology News, Solid State Technology, p. 22, May 1999.
[2.63] C.B. Labelle, K.K.S. Lau and K. Gleason, *Mat. Res. Soc. Symp. Proc.*, 511, 75 (1998).
[2.64] P.K. Wu, G.-R. Yang, X.F. Ma, and T.-M. Lu, *Applied Physics Letters*, 65, 508, 1994.
[2.65] P.K. Wu and T.-M. Lu, *Applied Physics Letters*, 71, 2710, 1997.
[2.66] S. Nitta, A. Jain, V. Pisupatti, W.N. Gill, P.C. Wayner, J.L. Plawksky, *Mat. Res. Soc. Symp. Proc.*, 511, 99 (1998)
[2.67] C.V. Nguyen, K.R. Carter, C.J. Hawker, J.L. Hendrick, R.L. Jaffe, R.D. Miller, J.F. Remenar, H.-W. Rhee, P.M. Rice, M.F. Toney, M. Trollsas, D.Y. Yoon, *Chem. Mater.*, **11(11)**, 3080 (1999).
[2.68] D. Mecerreyes, N. Kamber, E. Huang, V. Lee, T. Magbitang, W. Volksen, C.J. Hawker, R.D. Miller, J. L. Hedrick, *Polym. Prepr.*, **41(1)**, 517 (2000).

[2.69] E.M. Zielinski, S.W. Russell, R.S. List, A.M. Wilson, C.Jin, K.J. Newton, J.P. Lu, T. Hurd, W. Y. Hsu, V. Cordasco, M. Gopikanth, V. Korthuis, W.Lee, G. Cerny, N.M. Russell, P.B. Smith, S. O'Brien, and R. H. Havemann, *IEDM Tech. Digest*, December 7-10, 936 (1997).
[2.70] S.T. Chen, *Mat. Res. Soc. Symp. Proc.*, **381**, 141 (1995).
[2.71] K.R. Carter, et. al., *Mat. Res. Soc. Symp. Proc.*, **3891**, 79 (1995).
[2.72] J.F. Remenar, et. al., *Mat. Res. Soc. Symp. Proc.*, **511**, 69 (1998).
[2.73] S. Nitta, A. Jain, V. Pissupatti, W.N. Gill, P.C. Wayner, and J.L. Plawsky, *Mat. Res. Soc. Symp. Proc.*, **511**, 99-104 (1998).
[2.74] T. Ramos, et. al., *Mat. Res. Soc. Symp. Proc.* **511**, 105 (1998).
[2.75] A. Jain, S. Rogojevic, F. Wang, W.N. Gill, P.C. Wayner, J., and J.L. Plawsky, *Mat. Res. Soc. Symp. Proc.*, **512**, D5.25.1 (2000).
[2.76] S. Rogojevic, Ph.D. Thesis, Chem. Eng. Dept., Rensselaer Polytechnic Institiute, Troy, New York (2001).
[2.77] S. Rogojevic, A. Jain, F. Wang, W.N. Gill, P.C. Wayner, Jr., and J.L. Plawsky, *Journal of Vacuum Science and Technology*, **B19(2)**, 354-360 (2001).
[2.78] S. Rogojevic, A. Jain, W.N. Gill, and J. L. Plawsky, "Interactions Between Nanoporous Silica and Cu, J. Electro. Chem. Soc., In Press, submitted, (2002)
[2.79] A. Jain, S. Rogojevic, S. Pnoth, N. Agarwal, I. Matthew, W.N. Gill, P. Persans, M. Tomozawa, J.L. Plawsky, and E. Simonyi, *Thin Solid Films*, 398-399, 513-522 (2001).
[2.80] A.L.S. Loke, et., al., *IEEE Trans. Elect. Devices*, **46**, 2178 (1999).
[2.81] A. Jain, S. Rogojevic, W.N. Gill, J. Plawsky, I. Matthew, M. Tomazawa and E. Simoyi, *J. Appl. Phys.*, **90(11)**, 5832-5834 (2001).
[2.82] J. Baskan, et. al., *Adv. Mater.*, **12,** 291, (2001).
[2.83] C.J. Brinker and Scherer, "Sol Gel Science-the Physics and Chemistry of Sol Gel Processing", *Academic Press*, NY, (1990).
[2.84] A. Jain, S. Rogojevic, S. Pnoth, W.N. Gill, J.L. Plawsky, "Processing Dependent Thermal Conductivity in Nanoporous Silica Xerogel Films", *J. Applied Physics*, In Press (2002).
[2.85] S.S. Prakash, C.J. Brinker, and A.J. Hurd, *Journal of Non-Crystalline Solids*, **190**, 264-275 (1995).
[2.86] S.S. Prakash, C.J. Brinker, A.J. Hurd, and S.M. Rao, *Nature*, 374, 439-443 (March 1995).
[2.87] L.W. Hrubesh and S.R. Buckley, *Mat. Res. Soc. Symp. Proc.*, **476**, 99-110 (1997).

Chapter 3

# CHEMICAL-MECHANICAL PLANARIZATION (CMP)

Chemical-mechanical planarization (CMP) has found application in semiconductor processing as a method of controlling the planarity of the multiple metal and dielectric layers that form the IC interconnect structure [3.1, 3.2]. Nonplanarity is introduced to the wafer surface at the transistor isolation level and increases as the number of metal layers increases. CMP is the process of physically removing material from places of high topography to flatten and level the wafer surface. Figure 3.1 depicts nonplanarity increasing as metal layers increase, while Figure 3.2 depicts the ideal case of planar interconnect layers allowed by CMP.

When properly used, CMP can provide planar topography over devices, and extend this planarity uniformly across an entire die (or chip), or even across the wafer. Planarity across the die, or "global" planarity, is important for photolithography processes, which projects a pattern of light onto the wafer surface. If the surface is not planar, the focus of the lithography process will be incorrect, resulting in improper line width and spacing of interconnect structures. Planarity across the entire wafer is less critical in conventional ICs fabricated with stepper lithography. However, in fabrication processes using full-water lithography and more advanced wafer-scale three-dimensional (3D) ICs, full wafer planarity becomes important.

CMP is often considered to be a less clean, or even "dirty" process in the cleanroom fabrication environment. This is not a completely unfair label, since the process involves the face-down placement of the wafer into a slurry that contains millions of abrasive particles, each of which, in the normal cleanroom environment, would be considered a potential particle defect. For this reason, CMP tools are often housed in separate areas from the main fab, to reduce the possibility of particle migration from the CMP tools to particle-sensitive tools such as steppers and deposition chambers. A typical rotary-type manufacturing CMP tool [3.3] is shown in Figure 3.3.

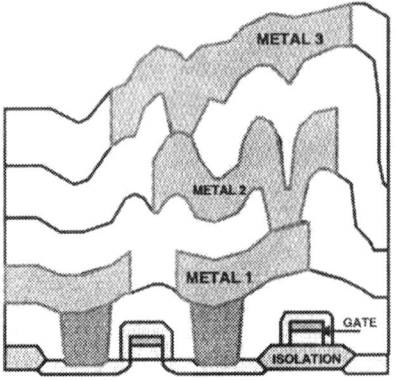

*Figure 3.1.* Schematic of nonplanar interconnect layers [3.2]

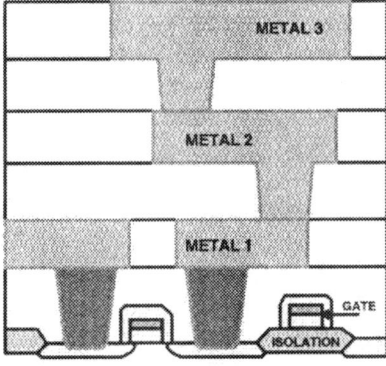

*Figure 3.2.* Schematic of the same interconnect layers structure, but with planar interconnect layers achieved with CMP [3.2]

Wafers are placed in a wafer cassette or front-opening unified pod (FOUP) into a wet or dry holding area within the CMP tool. The tool then selects the proper wafer based on the user input, removes the wafer from the cassette, places it in the polishing head, moves it to its proper position over the selected platen, and executes the programmed polishing process. After polishing in a stand-alone (dry-in/wet-out) tool configuration, the wafer is briefly cleaned using a water spray, and returned to a wafer cassette within a wet bath. In an integrated polish/clean tool (dry-in/dry-out), wafers move directly from the polish operation to a single-wafer cleaner/drier, and are then returned to their original dry cassette. In either configuration, it is very important that the polished wafers remain wet until post-CMP cleaning, since the wafers remain coated with a residue of slurry particles after CMP and slurry particles are extremely difficult to remove if allowed to dry on the water surface.

Post-CMP surface cleaning is a vital part of the CMP unit process [3.4]. Methods of cleaning include physical means such as brushcleaning [3.5] and non-contact means such as megasonic vibration cleaning within a wet chemical bath [3.6]. The chosen method must return the wafers to the high standards of cleanliness (extremely low particle levels) required for further cleanroom processing.

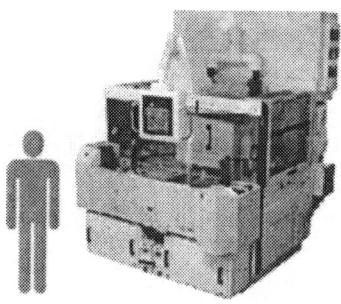

*Figure 3.3.* IPEC/Westech 472 double-platen CMP tool (from [3.3])

## 3.1 CMP PROCESS DESCRIPTION

The CMP process occurs when the semiconductor wafer is placed face down on the polishing pad and rotated. In the common rotary configuration pictured in Figure 3.3, the polishing pad adheres to a solid steel platen, which is rotated in the same direction as the wafer. As the wafer and pad rotate a pressure is applied to the back side of the wafer. A slurry that consists of chemical reagents and abrasive particles is dispensed to the center of the polishing pad, and spreads across the pad to create a lubricating layer between the pad surface and the wafer. A top-view schematic of the process is shown in Figure 3.4(a). The process results in removal of material from the wafer surface at a desired rate and with a desired surface finish.

Figure 3.4(b) shows a side view of the CMP process. The wafer is held in the wafer carrier by a retaining ring. The wafer protrudes slightly from this ring, ensuring the best opportunity for the wafer to contact the slurry and pad. Pressure is applied normal to the pad surface through the polishing arm to the wafer, slurry layer, and pad. A flexible carrier film or membrane is located behind the wafer, to help distribute the applied pressure to all points on the wafer surface. The wafer carrier is attached to the polishing arm by a gimbal point that allows the wafer carrier to tilt in the horizontal plane and achieve a position that supports the speed and pressure of the wafer. The slurry creates a layer between the pad and wafer for fluid shear, mechanical abrasion, and chemical reaction to occur.

Several of the parameters that control the CMP removal and finish of the material on the wafer surface are listed in Table 3.1. The wide array of parameters and the competing interaction of effects makes CMP a difficult process to benchmark, model, and predict.

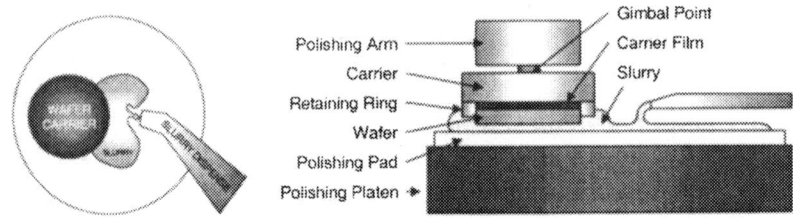

*Figure 3.4.* (a) Top view schematic of a single CMP platen; (b) side view schematic of a single CMP platen

*Table 3.1.* Adjustable CMP parameters.

| Parameter | CMP Effect |
|---|---|
| Pressure | Removal Rate, Removal Uniformity |
| Carrier Velocity | Removal Rate, Removal Uniformity |
| Pad Velocity | Removal Rate, Removal Uniformity |
| Slurry Chemistry | Removal Rate, Surface Topography |
| Slurry Abrasive (type, size, percent) | Removal Rate, Surface Topography |
| Pad Hardness | Removal Rate, Removal Uniformity, Surface Topography |
| Pad Porosity | Slurry Distribution, Removal Uniformity |
| Carrier Film Hardness | Removal Uniformity |

The name "chemical-mechanical planarization" alludes to the fundamental nature of the material removal process. CMP is designed to advantageously use chemical reaction and mechanical energy in combination. Chemistry and mechanics are combined to achieve remove material from high regions on the wafer surface, while leaving the low regions of the surface relatively untouched. This selective removal leads to surface planarization.

An enlarged view of the wafer/slurry/pad interface is shown in Figure 3.5, to illustrate the method by which chemistry and mechanics provide a planar post-CMP surface. The slurry contains chemicals that react with the wafer surface, and abrasive particles that impact the wafer surface to achieve mechanical removal. The highest areas of metal or dielectric on the wafer surface experience the highest shear stress and best chance for contact with the abrasives and polishing pad. This allows preferential removal of the high areas and overall planarization [3.7].

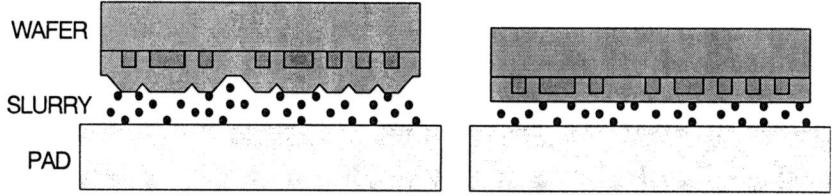

*Figure 3.5.* Schematic diagram depicting the combination of chemistry and mechanical action to planarize the wafer surface

The choice of slurry abrasive particle is vital for achieving the desired removal rate and surface roughness of a material as the result of a CMP process. Abrasives vary in size, shape, charge (isoelectric point in water), and mechanical hardness. Some typical CMP slurry abrasives and their isoelectric points in water are listed in Table 3.2 [3.8]. Cerium oxides ($CeO_2$), zirconium oxides ($ZrO_2$), and silicon oxides in high pH solutions ($SiO_2$) are commonly used to polish silicon dioxide, and aluminum oxide ($Al_2O_3$) is most commonly used for metal (Cu, W, Al) CMP.

*Table 3.2.* Potential CMP abrasive particles (from [3.8])

| Abrasive | Isoelectric Point | $SiO_2$ Polishing Rate ($\mu$m/min) |
|---|---|---|
| $CeO_2$ | | |
| $CeO_2$ (hydrous) | 6.8 | 4.9 |
| $\alpha$-$Fe_2O_3$ (hematite) | | |
| $Fe_2O_3$ (hydrous) | 8.5 | 3.1 |
| $TiO_2$ (rutile) | | |
| $TiO_2$ (hydrous) | 6.2 | 3.0 |
| $ZrO_2$ | | |
| $ZrO_2$ (hydrous) | 6.7 | 2.8 |
| $\alpha$-$Al_2O_3$ | 9.5 | 1.8 |
| $SiO_2$ | 2.2 | 0.01 |

## 3.2 CMP PROCESSES WITH COPPER METALLIZATION

While a full summary of CMP is beyond the scope of the book, the CMP processes incorporated in the most advanced IC interconnect structures are briefly summarized. These include oxide CMP required for planarization after device processing (front-end wafer processing) and before interconnect processing (back-end wafer processing), copper CMP and damascene patterning.

### 3.2.1 Oxide CMP

Polishing techniques have been used to polish glass, or silicon dioxide ($SiO_2$) for hundreds of years [3.9]. Forming optical lenses, polishing mirrors, and planarizing insulating layers on a semiconductor wafer surface are all examples of glass polishing. Silicon dioxide CMP occurs by both chemical and mechanical mechanisms. The mechanical mechanism is due to slurry flow past the $SiO_2$ surface causing abrasive impact and slurry shear erosion. The chemical mechanism exists in concert with the mechanical mechanism, as mechanical pressure and abrasion force the slurry chemicals into the $SiO_2$ surface [3.10-3.12]. There the chemicals react with the $SiO_2$ structure, weakening the surface bonds and allowing the material to be more easily abraded. The chemical reaction that is believed to exist in silicon dioxide CMP is as follows [3.10]:

$$\equiv Si - O - Si \equiv \ + H_2O \rightarrow 2 \equiv Si - OH \qquad (3.1)$$

Equation 3.1 is widely accepted as the chemical reaction that weakens the $SiO_2$ surface, increasing the effectiveness of the slurry mechanics. The complex nature of CMP is illustrated by the mechanism for CMP removal. To achieve material removal, individual chemical and mechanical effects contribute, plus a combined chemo-mechanical effect (slurry penetration into the glass surface).

Silicon dioxide CMP is one of the best understood CMP processes, since it has been studied for many years. For this reason, an abundance of data exists showing trends for the CMP removal rate of $SiO_2$ versus the various CMP parameters.

### 3.2.2 Copper CMP

The ability to select slurry chemistries and abrasive types makes CMP useful for the removal of many different materials. It is possible to take advantage of slurry chemistry to enable removal of metals that are very strong and resistant to abrasion. Metal CMP slurries may include strong etching and oxidizing chemistries to attack metals and dissolve them. Metal slurries may also include a passivating agent to protect the metal surface from etching in undesired areas.

Figure 3.6 illustrates how a combination of etch and passivation are combined to achieve removal. A passivating chemical in the slurry reacts with or adsorbs on the metal surface, preventing the chemical etching of the metal. During CMP, the slurry shear force and wafer contact with the pad and abrasive particles remove the passivating layer from the metal surface at the high regions, where shear is highest and contact is most likely. The exposed metal is then etched or oxidized by the slurry chemistry, planarizing the high regions to the level of the low regions.

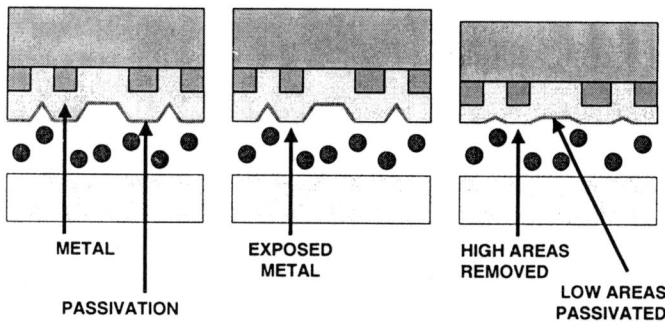

*Figure 3.6.* Schematic diagram showing the passivation / etch removal approach used for the CMP of metals.

Copper CMP may be performed in either acidic or basic media, by the formation of two different types of chemical surface layers. At high pH, with a slurry component such as $NH_4OH$, copper will react to form a hydrated copper layer [3.13]:

$$Cu + 2OH^- \rightarrow Cu(OH)_2 + 2e^- \qquad (3.2)$$

This altered layer is removed by the mechanical action of the pad and slurry abrasives. At high areas, the hydrated layer is removed rapidly, exposing fresh Cu. This mechanism continues, providing planarization and Cu removal.

Copper metal may also be removed in low-pH media by a passivation and etch CMP strategy. This involves a slurry chemistry that rapidly etches copper, such as nitric acid ($HNO_3$), and a slurry component that passivates the copper surface and blocks the etching reaction. Benzotriazole (BTA) is an organic compound that is known to adsorb on copper surfaces and protect against corrosion of copper in aqueous environments. Copper is normally etched in a solution with pH less than 7.0, as the copper dissolves as $Cu^{2+}$ ions while oxygen is reduced [3.14]:

$$Cu \rightarrow Cu^{2+} + 2e^- \qquad (3.3)$$
$$O_2 + 2H_2O + 4e^- \rightarrow 4OH^- \qquad (3.4)$$

But when BTA is present in the slurry, the copper is protected, inhibiting reactions 3.3 and 3.4:

$$Cu + BTA \rightarrow Cu - BTA \qquad (3.5)$$

The slurry shear, abrasive, and pad contact remove the BTA and allow the copper to etch and dissolve from high regions. BTA remains to protect the low regions, causing a difference in removal rate, or "selectivity" that allows planarization.

### 3.2.3 Damascene Patterning

Ancient metal workers used a technique known as damascene to create patterns of metal in the carved depressions of sword blades and handles. Patterns were made in the steel of the weapon, after which metal such as gold was pressed into the depression and polished to create an intricate design. Hundreds of years later the semiconductor industry has found use of a similar technique for fabricating patterns in chip interconnects which are thousands of times smaller than the swords of their ancient predecessors.

As mentioned in Chapter One, CMP is the instrumental technique that allows damascene copper patterning. International Business Machines (IBM) pioneered the development of both CMP and damascene patterning as enabling technologies for copper metallization [3.15]. Damascene patterning requires the proper physical CMP parameters such as velocity and pressure, and also the proper chemical slurry constituents, to create selectivity, or preferential removal of one material to another. Single damascene patterning consists of several simplified steps:

1) etch stop deposition
2) dielectric deposition
3) photoresist deposition and patterning
4) plasma etching of the dielectric material down to the etch stop layer
5) removal of remaining photoresist and etch stop layer
6) lining the trench or via pattern in the dielectric with a thin barrier
7) overfilling the trench or via with conductor metal
8) CMP to remove the metal overburden and the barrier material, stopping on the dielectric surface and leaving a planar interconnect layer
9) capping of the line or via

Figure 3.7 shows the metal overburden above via holes before CMP, and after CMP has been performed to remove the excess metal and define the circular contact holes. This process is known as "single damascene", where the via level and the trench or line level are patterned and polished in two completely different steps. In pre-copper interconnect technology, the via material is conventionally tungsten (W), and the line material was W or aluminum (Al). With the advent of copper metal, there has been a push towards patterning, filling, and polishing both the via level and the trench level in the same step. This process is known as "dual damascene" [3.16]. Single and dual damascene strategies are compared side by side in Figure 3.8. The dual damascene technique shown in the figure is a "trench first" patterning technique.

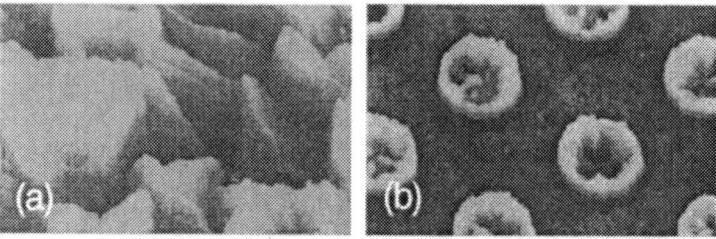

*Figure 3.7.* Tungsten vias (a) filled with metal before CMP, (b) after CMP (from [3.3])

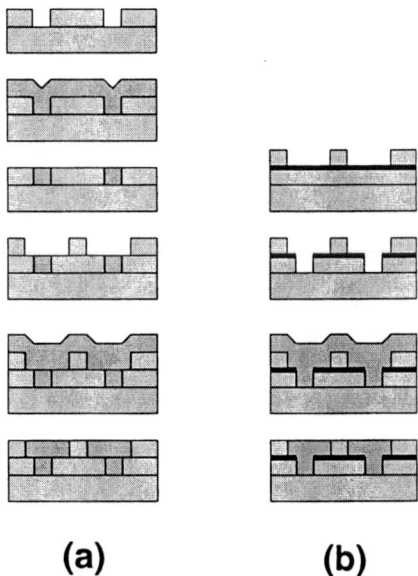

**(a)**          **(b)**

*Figure 3.8.* (a) Single level damascene patterning process flow for via and trench, and (b) dual damascene patterning process flow for via and trench)

Dual damascene (DD) copper technology often requires an additional etch stop material, the dark black layer between dielectric layers in Figure 3.8(b). This layer has a much lower plasma etch rate than the insulator material, causing the trench etch to stop at the desired depth. Then another layer of photoresist is coated and patterned to produce the via openings in the dielectric. The etch stop, commonly a silicon nitride material ($SiN_x$, κ ~ 7.0) remains between the dielectric layers, causing an increased effective dielectric constant. There are many variations on dual damascene, with the most popular being "trench first", "via first", and "buried etch stop" [3.17] [3.18]. Each variation has its own processing difficulties [3.19], mainly caused by the higher aspect ratio of the etching dimensions and the post-etch (prior to metal deposition) cleaning of residues deep within the trench/via [3.20].

One advantage of CMP and DD patterning has been discussed previously -- the ability of CMP to provide planar surfaces for better photolithography yield over multiple interconnect levels. Other advantages of CMP are summarized in Table 3.5. Aside from planarity advances, perhaps the other most useful aspect of CMP is illustrated in points (2) and (3). Multilayer stacks require planarization of metal, liner, and several types of dielectrics. CMP can be used to mechanically or chemically remove a multitude of materials, making it extremely useful for interconnect fabrication. The other advantage of note is that the damascene CMP process increases device yield and reliability. This is due to the fact that CMP eliminates etches and alternative planarization methods that require multiple clean and handling steps [3.21]. The reduction of processing steps is reflected in the overall device yields.

*Table 3.5.* Advantages of Damascene Patterning (from [3.21])

| |
|---|
| (1) Achieves global planarization |
| (2) Universal or materials insensitive – all types of surfaces can be planarized |
| (3) Useful for multi-material surfaces |
| (4) Reduces severe topography allowing for fabrication with tighter design rules for additional interconnection levels |
| (5) Provides an alternative means of patterning metal, eliminating the need of the reactive ion etching or plasma etching for difficult-to-etch metals and alloys |
| (6) Leads to improved metal step coverage (or equivalent) |
| (7) Helps in increasing reliability, speed, and yield (lower defect density) for sub-0.5 μm devices/circuits |
| (8) A potentially low cost process |
| (9) Does not use hazardous gases in dry etching processes |

Whether single or dual damascene patterning is used to fabricate interconnect structures, the interaction of the CMP consumables with the wafer surface will have a large impact on the final device yield. The pad and slurry mechanically and chemically alter the wafer surface, so performance metrics regarding the damascene CMP process (such as removal rate, surface scratching, etc.) are desired to guide process development.

### 3.2.4 CMP Targeted Results and Challenges

There are several metrics that must be evaluated when performing a CMP process in a manufacturing environment. These metrics fall under "desired" and "undesired" categories – with the goal being to achieve as many of the desired results as possible, while causing as few of the undesired results to occur. As many IC fabrication unit processes, CMP must be a balance between these factors.

The main "desired" metrics are material removal and wafer planarization. The goal of any CMP process is to remove material from the wafer surface. In the case of gap-fill planarization, the goal is to remove the overlaying dielectric, resulting in a planar interconnect level, as shown in Figure 3.6. In the case of damascene CMP, the goal is to remove the metal overburden and liner material from the wafer field (as shown in Figure 3.8) while keeping metal within the trench or via. Each CMP process would optimally result in a perfectly planar wafer surface, although differing interconnect densities, widths, and spacings make this difficult to achieve [3.22, 3.23].

The main "undesired" metrics are metal or dielectric scratching [3.24], metal or dielectric corrosion or penetration with chemicals [3.25], excessive metal removal from within trenches and vias (dishing) [3.26], excessive dielectric removal in dense arrays (erosion) [3.26], and post-CMP slurry residual particulates [3.27]. The main challenge of CMP is to achieve material removal in a controlled fashion – without deep scratching and gouging, and without rapid chemical etching. It is very important to achieve a balance between chemical and mechanical effects, as shown in Figure 3.8. It is most desirable to chemically alter the material surface to a form that is more easily physically removed, and then abrade/remove this altered layer using mechanical force. This synergism represents the most difficult challenge in CMP processing, both to achieve in a manufacturing environment and to fundamentally understand.

## 3.3 CMP OF LOW-κ MATERIALS

It is common practice for researchers to investigate the CMP process of a new material by using a slurry chemistry that has been well defined for a different well-known material. For example, the CMP of FSG has been compared to the CMP of CVD TEOS oxide films in oxide slurries. In one slurry developed for oxide polish, FSG is removed at a rate approximately 10 percent higher than undoped TEOS oxide [3.28]. This increase is due to the bonding structure of FSG, where fluorine incorporation causes termination in the silica structure, leading to lower hardness and elastic modulus [3.29]. This is a common problem with low-κ materials, since changes in the bonding structure that reduce electrical polarization and dielectric constant usually result in reduced film hardness. In addition, for a given slurry concentration, removal rate increases greatly with the fluorine content of the film, as seen in Figure 3.9. This result matches previous understanding, and requires the proper selection of fluorination to achieve desired dielectric constant but retain polishing properties similar to TEOS oxide.

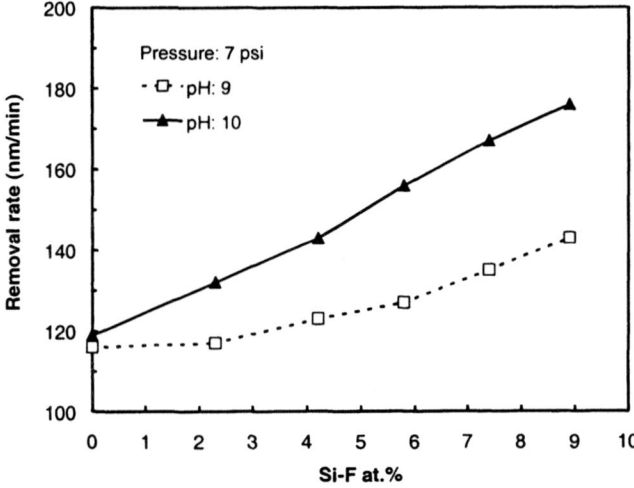

*Figure 3.9* – The CMP removal rates of SiOF films with different fluorine concentrations under a constant pressure of 7 psi (from [3.29]).

The fluorine in the FSG also reduces the chemical resistance of the oxide films. Researchers have observed an increase in the FSG refractive index (RI) after CMP, which has been attributed to moisture absorption during CMP. Capping the FSG with a TEOS oxide layer can protect the surface from slurry chemical attack [3.30]. The cap layer provides sufficient adhesion and eliminates the chemical penetration into the FSG, but the use of TEOS oxide (κ ~ 3.9) cap results in a higher effective dielectric constant.

The family of carbon-doped silica (SOG, HSQ, MSQ, OSG) has been tested for CMP processing in a manner similar to FSG. Slurries and CMP parameters used for oxide CMP have been used to polish carbon-doped glasses for both gap-fill [3.31] and damascene interconnect formation [3.32]. Previous work has shown that the CMP removal rate for oxide/organic spin-on glasses (SOG) varies with the organic content and thermal curing of the films [3.33] when performing CMP with slurries developed for SiO$_2$ polishing. The organic content of the film suppresses the SOG removal rate below that of plasma deposited silicon dioxide due to the hydrophobicity of the oxide/organic films, as shown in Figure 3.10. However, the lower temperature cure used to crosslink some SOG materials may result in high film removal rate and damage.

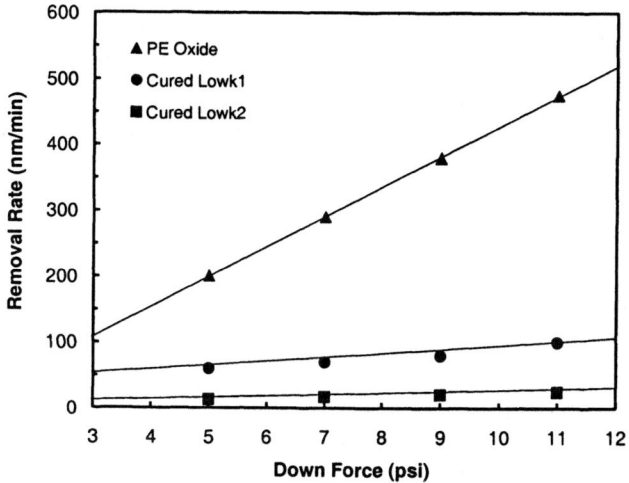

*Figure 3.10* – The relationship of CMP removal rate and downforce for two low-κ SOGs and PECVD oxide (from [3.33])

Others have compared the direct CMP of organosilicate poly(silsesquioxanes) to a fluorinated aryl-ether polymer for planarization of gap-fill structures [3.34]. The CMP of OSG silsesquioxanes was found to be dependent on slurry pH, with very smooth post-CMP surfaces possible. Further work with silsesquioxanes examined overcoming the film hydrophobicity, which suppresses removal rate, by selecting slurry chemistry and additives to decrease the film surface energy [3.35]. Figure 3.11 shows CMP results for HSQ and MSQ films polished using SS25 slurry from Cabot Corporation, with or without tetramethylammonium hydroxide (TMAH) surfactant. The surfactant enhances the surface wetting of HSQ and MSQ, increases the contact area for slurry/wafer interaction and results in higher observed CMP rate. Slurry wetting and surface hydrophobicity are important topics in low-κ CMP, as the materials focus shifts towards more carbonaceous and hydrophobic films. The CMP of OSGs is extensively presented in Chapter 5.

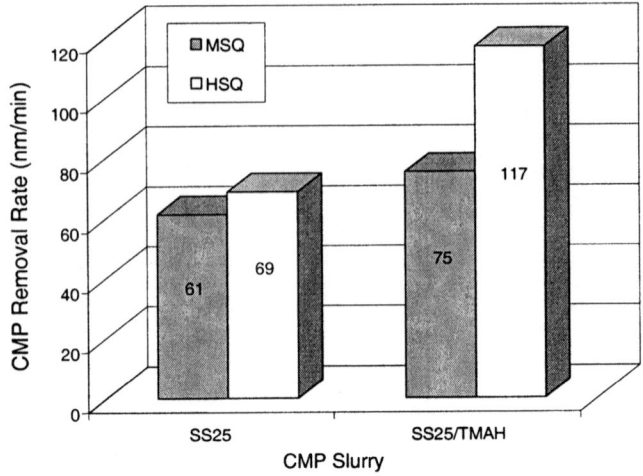

*Figure 3.11* – Variation of removal rate of hydrogen- and methyl- silsesquioxanes using SS25 slurry with and without TMAH surfactant additive (from [3.35]).

Early investigations into the CMP of low-κ polymers compare the removal rate of a crosslinked thermoset (benzocyclobutene or BCB) and two linear-chain thermoplastics (parylene-n and fluorinated polyimide) [3.36]. Slurries developed for copper CMP containing alumina ($Al_2O_3$) abrasive particles with pH ranging from 0.5 to 9.5 removed each polymer with rate close to 1000 nm/min. Results obtained using small polymer-coated wafer pieces and a metallographer's wheel are shown in Figure 3.12.

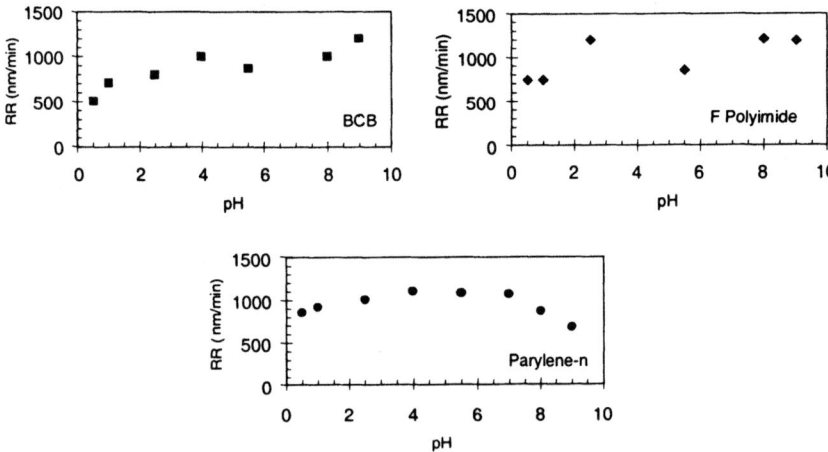

*Figure 3.12* – CMP removal rates for BCB, fluorinated polyimide, and Parylene-n using copper slurries (from [3.36])

All three polymers have high removal rates, with high scratching and post-CMP defect levels suggesting that the polymers were mechanically abraded. Increased BCB removal rates are also observed with lower cure temperature and time, suggesting that polymer hardness affects the mechanical impact of the CMP slurry. Adding surface-active agents (surfactants) to CMP slurries also allows higher BCB removal rate [3.37]. The nonionic surfactant, Triton-X 100, improves the wetting of the hydrophobic polymer by the aqueous slurry. Wetting agents have been used for some time in applications requiring the wetting and dissolution of polymers in aqueous solutions. The surfactant molecules appear to adsorb on the polymer, in some cases causing swelling of the polymer structure, and in all cases allowing better suspension of polymer particles in aqueous media. Beyond a threshold surfactant concentration required to enable removal, the BCB removal rate is constant regardless of increased surfactant concentration in the CMP slurry, as shown in Figure 3.13. The CMP of BCB is more extensively described in Chapter 4.

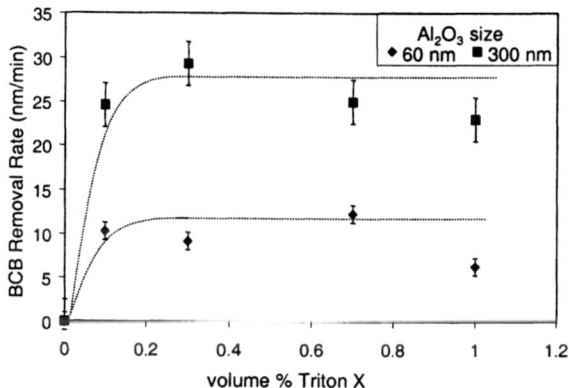

*Figure 3.13* – BCB removal rate versus amount of Triton-X for two abrasive sizes. Slurries consist of 0.5 vol% $HNO_3$, 0.1 wt% $Al_2O_3$ abrasive, and surfactant in DI water (from [3.37])

Each surfactant has a characteristic critical micelle concentration (CMC), above which the surfactant molecules in aqueous media align to form large molecular weight clusters. These clusters can surround abraded polymer, suspending the polymer in the slurry and allowing transport of the polymer away from the film surface during CMP. These micelles, which form at the surfactant's CMC, are responsible for the plateau in removal rate.

The CMP removal rate and uniformity of polyimide in copper slurries may also be improved using organic slurry additives [3.38]. The addition of 20 – 30 vol% glycerol to acidic slurries containing alumina abrasive causes an increase in the CMP removal rate (see Figure 3.14). As the glycerol concentration is increased to 40 – 60 vol%, the CMP rate decreases the CMP rate due to an increase in the slurry viscosity. In addition, the uniformity of the polyimide film following CMP improves with added glycerol, while damage to the film surface decreases. The glycerol behaves like a surfactant, adsorbing on the polyimide surface and aiding in suspension of abraded material. Glycerol additive also increases the uniformity of copper polished in the same slurries. A similar enhancement in polyimide removal rate has been reported using TMAH surfactant additive, again illustrating the importance of surface hydrophobicity on polymer CMP [3.34].

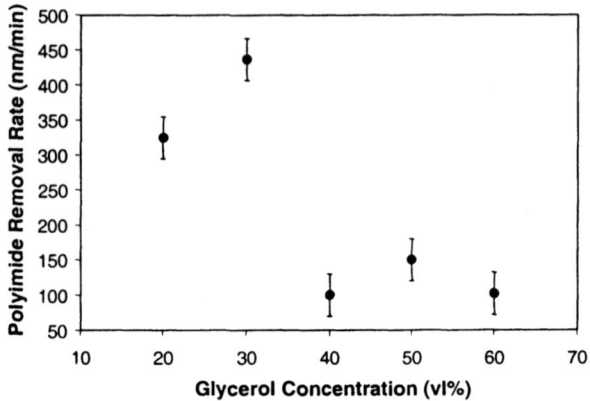

*Figure 3.14* – Polyimide polish rate vs. glycerol concentration. Polished on Strasbaugh, 50 rpm (~ 200 cm/sec), 5 psi (~ 34 KPa), slurry of 1 wt% $Al_2O_3$, pH =2 (from [3.38]).

Less successful attempts have been made to attain a non-damaging removal process for parylene-n [3.40]. The deposition mechanism for parylene-n results in long linear polymer chains with some crystallinity, but minimal crosslinking. Parylene-n is very hydrophobic due to the lack of oxygen or silicon in the polymer chain, but also has low hardness and shear strength. All of these factors have made efforts to control the direct CMP of parylene-n very difficult [3.41]. CMP in a variety of acidic and basic slurries, with and without surfactant additives, results in extensive damage to the polymer in the form of surface scratching.

Successful CMP of FLARE™ low-κ organic films has been shown through an extensive study of slurry chemistry, abrasive particles, and CMP pressures [3.42]. FLARE is a poly(arylene) ether spin-on polymer with high thermal stability (400 °C). Certain chemically active abrasives, such as zirconium oxide and cerium oxide remove FLARE at much higher rates than silica alternatives, as shown in Figure 3.15. The addition of ferric nitrate

---

™ - FLARE is a trademark of AlliedSignal, Inc.

oxidizer also greatly increases the CMP removal rate of FLARE, as shown in Figure 3.16.

The effects of chemical components in the abrasive and slurry suggest that the CMP of some polymers can be controlled using chemistries tailored to the polymer material. For example, a strong oxidizing chemistry accelerates the CMP removal rate of FLARE, a phenomenon that is not observed with BCB, parylene-n, or polyimide films. The specific chemistry suggested for FLARE removal is oxidation/reduction chemistry.

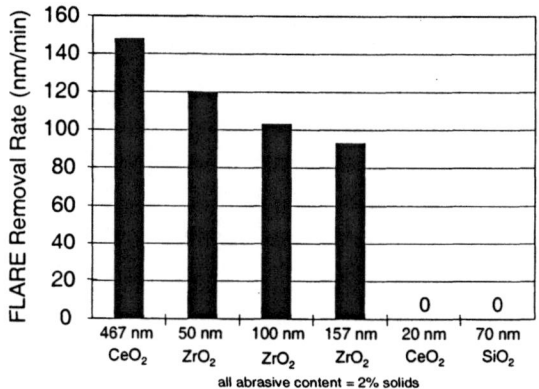

*Figure 3.15* – Removal rate of FLARE vs. selected specialized abrasives [3.42]

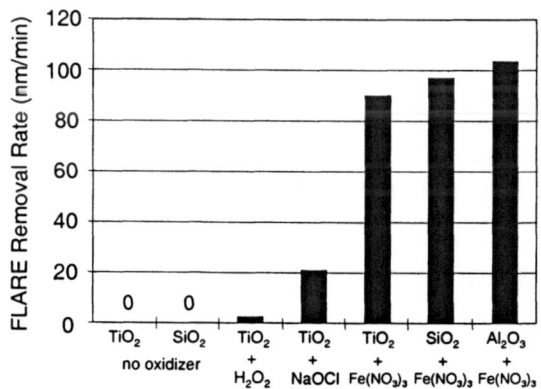

*Figure 3.16* – Removal rate of FLARE vs. various abrasives/oxidizers [3.42]

SiLK CMP has also been studied in a variety of slurry chemistries [3.43], [3.44] prior to our research. The chemical analysis plot in Figure 3.17 shows that CMP chemically alters the SiLK surface. This suggests that there is adsorption and chemical reaction of slurry components on the SiLK surface during CMP, which result in high (~ 400 nm/min) removal rates. The CMP of SiLK is extensively described in Chapter 4.

The direct CMP of porous materials has not been reported extensively in the literature to date. Conventional CMP may be more difficult for forming interconnects with copper and porous low-κ, due to reduced film adhesive and cohesive strength and the stress developed during CMP. However, porous materials are expected to react by the same mechanisms as dense materials, with additional enhancement in removal rate due to porosity, reduced material strength, and increased surface area for chemical reactions to occur. Thus silica xerogels, for example, are expected to be removed by the reaction in Equation 3.1, with the removal rate greatly accelerated due to the high porosity of the films. The same will be the case for porous polymers or silsesquioxanes. While conventional CMP may be more robust than initially considered, use of reactive liquids (abrasive-free CMP) or alternative planarization techniques like electropolishing are also being pursued.

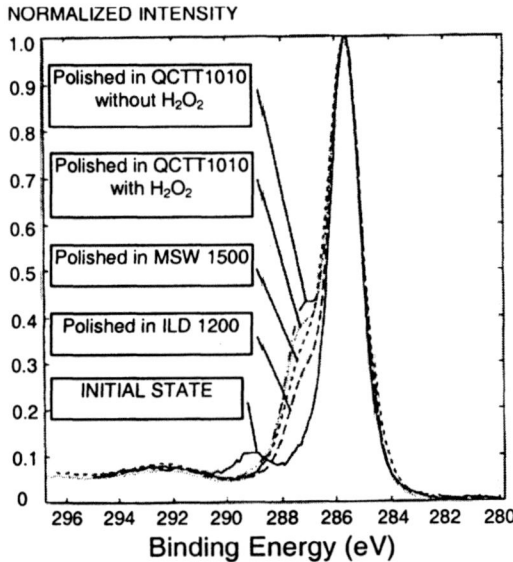

*Figure 3.17* – C 1s spectrum (from XPS) of SiLK after polishing using commercially available slurries (from [3.44])

## 3.4 CMP PROCESS MODELS

A brief overview of the models developed for CMP can provide some insight into the difficulties of fundamentally predicting removal rates, surface damage, and profile evolution during the CMP process. One of the first and simplest models for CMP was developed by Preston to relate the relative velocity ($V$), pressure ($P$), and slurry chemistry to $RR$, the material removal rate [3.45].

$$RR = kPV \qquad (3.6)$$

Preston's equation includes all slurry and pad effects in the empirical constant $k$. Its prediction of first order dependence on pressure and relative velocity is a good approximation, and is the starting point of the majority of the mechanical CMP models in the literature. However, the CMP process for most materials is better predicted by a $P^{2/3}$ dependence rather than $P$ [3.46]. The $P^{2/3}$ dependence can be predicted using fluid mechanics equations within the geometry of the three-dimensional CMP process. The fact that this dependence better matches the experimental data brings to light the fact that more complex models, based on fluid mechanics and contact mechanics, are required to capture the fundamental aspects of the CMP process -- at least at the wafer-scale. The polishing process for microelectronics is complex because it exists on multiple scales, from the "wafer-scale" (100s of mm) to the damascene "feature-scale" (less than a micron), a distance scale factor of more than five orders of magnitude. This discussion is related to wafer-scale models which predict the removal rate, slurry properties, and system geometry at the wafer scale.

### 3.4.1 Contact Mechanics-Based Models

As discussed in section 3.2, the wafer is pressed into the polishing pad, while a liquid slurry is applied at the leading edge of the wafer. The result is that the wafer surface will be exposed to a mixture of physical contact with the pad and abrasive and flow contact with the liquid slurry. A common approach to the CMP process is to treat the process as either a "flow" process or a "contact" process, and treat that single case [3.47].

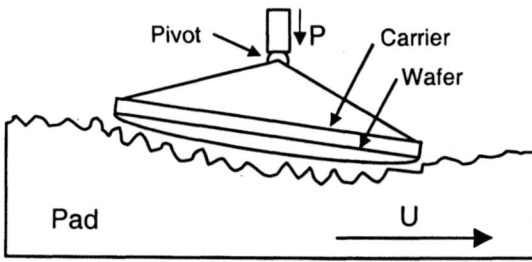

*Figure 3.18* – Schematic representation of a wafer polished based on pad contact mechanics (from [3.46]).

One of the first models that treated the wafer and pad surface to be in intimate contact was developed by Runnels and Renteln [3.48]. This model assumes that the wafer and pad are in intimate contact, as would be the case for high pressure and low velocity, as shown in Figure 3.18. The effect of motion and slurry are neglected, and the system is represented as a wafer being pressed against a stationary pad. A finite element solution calculates the pressure profile along the wafer, and predicts a maximum in the pressure 1 – 5 millimeters from the wafer edge, due to the deflection of the pad and wafer. Wang and Cale have extended this work to include the properties of the wafer, carrier, and carrier film to calculate the Von Mises stress at the wafer surface [3.49]. This treatment also predicts a high contact pressure near the wafer edge, as the result of compression and radial expansion of the wafer.

Luo and Dornfield have focused on the mechanical abrasion of particles contained with the pad asperities in establishing a solid-solid contact-based removal model. The model focuses on the volume of material removed by a single abrasive particle, assuming plastic contact over wafer-abrasive and pad-abrasive interfaces. A normal distribution of abrasive particle size and a period roughness of the pad surface are assumed. Although the chemical effects in CMP are introduced with a first-order fitting parameter, Luo and Dornfield have modeled mechanical abrasion and pad effects in an effective fashion which will be the basis for further modeling and simulation of this crucial component of the CMP process [3.51-3.53].

More recently, a model developed by Tichy and Levert approaches the contact mechanics of the CMP process and fluid mechanics under the wafer together [3.50]. The model represents the polishing pad as a set of springs, between which the slurry must flow. This is one of the few CMP models to attempt to describe the synergism that exists during CMP between mechanical removal and fluid flow. Both aspects of the problem should be addressed together, but the complexity of the CMP process generally causes fundamental treatment of one case or the other.

### 3.4.2 Fluid Mechanics-Based Models

Experiments have measured a liquid slurry film under the wafer surface to be 30 – 40 microns thick for a reasonable combination of pressure and velocity [3.54]. Depending on the height of the imperfections or "asperities" on the pad surface, this fluid may create a lubricating layer that can be described by fluid mechanics equations in three dimensions, such as the Navier-Stokes equation. One of the first fundamental models based on this premise was developed by Runnels and Eyman [3.55]. This is a wafer-scale model that predicts fluid flow, shear, and pressure gradient within the slurry layer. The geometry of the solution requires that the wafer must have a domed curvature and be tilted upward in the direction of the slurry flow (gap is wider at slurry inlet) to balance the force moments of the pressure and the relative velocity (see Figure 3.19). The model uses simplified Navier-Stokes fluid mechanics equations to solve the three dimensional pressure profile and thickness of the slurry film between the wafer and pad. Rogers, et al extended this fluid flow model to predict the slurry film thickness, velocity, and pressure profiles under the wafer surface, and also at the edges of the wafer carrier, extending out into the freestanding, unconstrained slurry film [3.56]. They used this model to compare the effects of pads on the flow or stagnation of the slurry under the wafer surface. Another model by Runnels [3.57] predicts feature scale removal rates and erosion profiles based on the pressure and velocity of the slurry film using a shear based erosion formula. Both models account for the physics of fluid mechanics, but treat chemical effects empirically.

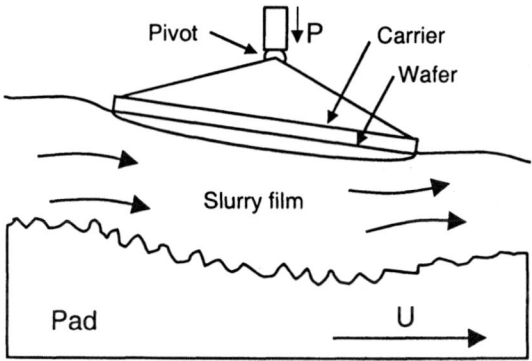

*Figure 3.19* – Schematic representation of Runnels' wafer scale tribological CMP model (from [3.55])

Sundararajan [3.58] uses Runnels' geometry and couples fluid mechanics, chemical reactions, and fundamental mass transfer mechanisms to attain a mechanism for copper CMP. This model is the basis for Thakurta's treatment of the CMP process, which incorporates pad properties (porosity, flexibility) into the calculation [3.59]. This type of model, which couples fluid mechanics, mass transfer, and chemical reaction mechanisms at the wafer surface, is one of the most complete fundamental treatments of the CMP process, since it takes into account the fundamental fluid flow (shear stress, dissolution) and chemical reaction aspects of CMP.

Paul has done parallel work modeling tungsten CMP, also focusing on slurry chemistry and reaction rate modeling [3.60, 3.61]. Explicitly accounting for both dissolution and abrasion, Paul empirically establishes rate parameters that fit removal rate information over a wide range of pressure, rotational speed and oxidizer concentration.

The fluid mechanics-based models generally include a two-step removal process, namely chemical modification of the film surface layer followed by abrasion of this modified surface layer. The first-order universality of such a two-step model has recently been discussed [3.62]. While direct abrasion of the film surface and chemical dissolution are possible and can occur, such direct film removal is usually undesirable in an IC interconnect CMP process.

## 3.5 LANGMUIR-HINSHELWOOD SURFACE KINETICS IN CMP MODELING

The purpose of the various models discussed in section 3.4 is to represent and predict the behavior of the CMP process using fundamental mathematical equations. In order to do this, one must capture the essence of several parts of the problem. The CMP process, as we define it, consists of three parts: (1) slurry flow hydrodynamics, (2) mass transport of diffusive species to and from the wafer surface, and (3) chemical reaction of the slurry chemical with the wafer surface by one of the mechanisms described in section 3.2.2. Chemical reaction surface kinetics can be modeled simply as a proportional forward reaction, or as a complex multi-step reaction process [3.63].

This work features a multi-step surface reaction at the wafer surface during CMP. The formulation is constructed according to the basic principles of heterogeneous catalysis, using Langmuir-Hinshelwood (L-H) surface kinetics [3.64]. The L-H approach is as follows:
  1) assume a sequence of steps in the reaction
  2) choose between individual steps and mechanisms, such as adsorption, reaction, and desorption.
  3) write mathematical rate laws for each individual step, assuming that all steps are reversible
  4) assume one step to be rate-limiting, and use the other "equilibrium" steps to reduce terms

This approach has not been used previously to model the CMP process, but has found application elsewhere in the semiconductor industry. For example, the reactive etching of $SiO_2$ has been described by Roosmalen using L-H kinetics [3.65]. L-H kinetics has also been used to model the growth and etching of germanium films by CVD [3.66]. Both growth and etching occur by similar multi-step reaction schemes. One case results in a "catalyst building", and the other a "catalyst consumption". For both $SiO_2$ and Ge, film etching occurs by a four step process:
  1) diffusion of reactive plasma to the wafer surface
  2) dissociation of reactive plasma gasses

3) reaction of dissociated ions with the oxide surface to produce volatile products
4) diffusion of volatile products away from the wafer surface

In typical heterogeneous catalysis, gas phase molecules travel to a catalytic surface and become bound to catalyst surface sites. Neighboring molecules react, forming products on the catalyst surface. The products then desorb, leaving behind a vacant catalytic site. The catalyst does not take part in the chemical reaction at the surface, other than to provide neighboring sites at which two bound molecules react [3.67]. Each of these steps may be represented mathematically and solved by selecting one step to be slow, or rate determining. The result is an expression for the reactive plasma etch rate of $SiO_2$, as a function of rate constants and reactant and product concentrations.

The CMP process that we propose differs from typical heterogeneous catalysis in that the wafer surface, where the reactants adsorb, becomes involved in the chemical reaction. The surface itself reacts to form an altered layer, which then desorbs due to the slurry flow and contact with the pad and abrasive. The result is a "catalyst consumption" process that results in the reduction in film thickness that is measured as the CMP removal rate.

A fundamental model such as the L-H model for surface kinetics can provide valuable information about the mechanisms that occur during the CMP of various materials. Models, when coupled with experiments, can provide suggestions for optimal operating parameters such as those listed in Table 3.1. The purpose of each model is to gain knowledge that can be used to predict the CMP of current and new materials. This L-H based model is more fully described in Chapter 6.

## 3.6 REFERENCES

[3.1] J.M. Steigerwald, S.P. Murarka and R.J. Gutmann, Chemical-Mechanical Planarization of Microelectronic Materials, Wiley Interscience (1997).
[3.2] S. Wolf and R. N. Tauber, Silicon Processing for the VLSI Era, 2nd Edition, CA: Lattice Press (1999).
[3.3] http://www.ipec.com/cmp/products.htm
[3.4] F. Zhang, A. Busnaina, *Appl. Phys. A,* **69**, 437 (1999).
[3.5] G. Zhang, G. Burdick, F. Dai, T. Bibby, S. Beaudoin, *Thin Solid Films,* **332(1-2)**, 379 (1998).
[3.6] M. Eissa, S. Joshi, G. Shinn, S. Rafie, B. Fraser, *Proc. Electrochem. Soc.,* **99-37**, 499 (2000).
[3.7] J. Warnock, *J. Electrochem. Soc.,* **138(8)**, 2398 (1991).
[3.8] J. G. Darab, D. W. Matson, *J. Elect. Mat.* 27(10), 1068 (1998).
[3.9] J. M. Steigerwald, S. P. Murarka, R. J. Gutmann, Chemical Mechanical Planarization of Microelectronic Materials, New York: J. Wiley & Sons (1997).
[3.10] L. Cook, *J. Non-Cryst. Solids,* **120**, 152 (1990).
[3.11] J. A. Trogolo, K. Rajan, *J. Mat. Sci.,* 29, 4554 (1994).

[3.12] A. Agarwal, M. Tomozawa, W. A. Lanford, *J. Non-Cryst. Solids*, **167(1-2)**, 139 (1994).
[3.13] J. M. Steigerwald, S. P. Murarka, R. J. Gutmann, D. J. Duquette, *J. Electrochem. Soc.* **141(12)**, 3512 (1994).
[3.14] J. M. Steigerwald, S. P. Murarka, D. J. Duquette, R. J. Gutmann, in *Advanced Metallization and Interconnect Systems for ULSI Applications in 1994*, 173 (1995).
[3.15] P. C. Andricacos, C. Uzoh, J. O. Dukovic, J. Horkans, H. Deligianni, *Proc. Electrochem. Soc.*, **98-6**, 48 (1999).
[3.16] C. W. Kaanta, S. G. Bombardier, W. J. Cote, W. R. Hill, G. Kerszykowski, H. S. Landis, D. J. Poindexter, C. W. Pollard, G. H. Ross, J. G. Ryan, S. Wolff, J. E. Cronin, $8^{th}$ *Intl. VLSI Mult. Interconn. Conf. (V-MIC)*, June 11-12, Santa Clara, CA 144 (1991).
[3.17] D. T. Price, R. J. Gutmann, S. P. Murarka, *Thin Solid Films*, **308-309**, 523 (1997).
[3.18] D. T. Price, Ph.D. Thesis, Dual Damascene Copper Interconnects with Low-k Polymer Dielectrics, Rensselaer Polytechnic Institute, Troy, NY (1999).
[3.19] P. K. K. Ho, M.-S. Zhou, S. Gupta, R. Chockalingam, J. Li, M. Fan, *Proc. SPIE Int. Soc. Opt. Eng.*, **3883**, 34 (1999).
[3.20] A. Skumanich, M.-P. Cai, *Proc. SPIE Int. Soc. Opt. Eng.*, **3883**, 96 (1999).
[3.21] K. Leitner, N. Elbel, K. Koller, *Proc. SPIE Int. Soc. Opt. Eng.*, **3214**, 58 (1997).
[3.22] J. Xie, K. Rafftesaeth, S.-F. Huang, J. Jensen, R. Nagahara, P. Parimi, B. Ho, *Mater. Res. Soc. Symp. Proc.*, **566**, 223 (2000).
[3.23] T. Park, T. Tugbawa, D. Boning, S. Hymes, P. Lefevre, T. Brown, K. Smekalin, G. Schwartz, *Proc. Electrochem. Soc.*, **99-37**, 94 (2000).
[3.24] L. Zhong, J. Yang, K. Holland, J. Grillaert, K. Devriend, N. Heylen, M. Meuris, *Jpn. J. Appl. Phys. Pt 1*, **38(4A)**, 1932 (1999).
[3.25] H. Fusstetter, A. Schnegg, D. Graf, H. Kirschner, M. Brohl, P. Wagner, *Mater. Res. Soc. Symp. Proc.*, **386**, 97 (1995).
[3.26] L. M. Ge, D. J. Dawson, T. Cunningham, *Proc. Electrochem. Soc.*, **99-9**, 238 (1999).
[3.27] L. Zhang, S. Raghavan, M. Weling, *J. Vac. Sci. Technol. B*, **17(5)**, 2248 (1999).
[3.28] D. Mordo, I. Goswami, I. J. Malik, T. Mallon, R. Emami, B. Withers, *Mater. Res. Soc. Symp. Proc.*, **443**, 127 (1997).
[3.29] W.-T. Tseng, Y.-T. Hsieh, C.-F. Lin, M.-S. Tsai, M.-S. Feng, *J. Electrochem. Soc.*, **144(3)**, 1100 (1997).
[3.30] H. M'saad, M. Vellaikal, L. Zhang, D. Witty, *Mater. Res. Soc. Symp. Proc.*, **564**, 443 (1999).
[3.31] P. Sermon, K. Beekman, S. McClatchie, *Vacuum Solutions*, **5**, 31 (1999).
[3.32] I. Lou, P. Lee, J. Ma, T. Poon, C.I. Lang, D. Sugiarto, W. F. Yau, D. Cheung, S. Li, B. Brown, X. Li, M. Naik, *16th Intl. VLSI Mult. Interconn. Conf. (V-MIC)*, September 7-9, Santa Clara, CA 234 (1999).
[3.33] S.-Y. Shih, L.-J. Chen, *3rd Intl. Conf. on CMP Planar. (CMP-MIC) Conference*, February 19-20, Santa Clara, CA 305 (1999).
[3.34] L. Forester, D. K. Choi, R. Hosseini, United States Patent No. 5,952,243, Issued September 1999.
[3.35] W.-C. Chen, C.-T. Yen, *J. Vac. Sci. Technol. B*, **18(1)**, 201 (2000),
[3.36] J.M. Neirynck, S.P. Murarka, R.J. Gutmann, in: T.-M. Lu, S.P. Murarka, T.-S. Kuan, C.H. Ting, *Low-Dielectric Constant Materials - Synthesis and Applications in Microelectronics*, San Francisco, USA, April 17-19, 1995, *Mat. Res. Soc. Symp. Proc.*, **381**, 229 (1995).
[3.37] J.M. Neirynck, G.-R. Yang, S.P. Murarka, R.J. Gutmann, *Thin Solid Films*, **290-291**, 447 (1996).

[3.38] D. Permana, S.P. Murarka, M.G. Lee, S. I. Beilin in: R. Havemann, J. Schmitz, H. Komiyama, K. Tsubouchi, *Advanced Metallization and Interconnect Systems for ULSI Applications in 1996*, Boston, USA, October 1-3, 1996, Proceedings of Advanced Metallization and Interconnect Systems for ULSI Applications in 1996, 539 (1997).

[3.39] Y.L. Tai, Y.L. Wang, *$4^{th}$ Intl. Conf. on CMP Planar. (CMP-MIC) Conference*, March 2-3, Santa Clara, CA 379 (2000).

[3.40] S. Rogojevic, J.A. Moore, W.N. Gill, *J. Vac. Sci. Technol. A*, **17(1)**, 266 (1999).

[3.41] G.-R. Yang, Y.-P. Zhao, J. M. Neirynck, S. P. Murarka, R. J. Gutmann, *J. Electrochem. Soc.*, **144(9)**, 3249 (1997).

[3.42] D. Towery and M. Fury, *J. Elect. Mat.*, **27(10)**, 1088 (1998).

[3.43] F. Küchenmeister, Z. Stavreva, U. Schubert, K. Richter, C. Wenzel, Adv. Metal. Conf., Colorado Springs, CO, (1998).

[3.44] F. Küchenmeister, U. Schubert, J. Heeg, C. Wenzel, Proc. of the Intl. Interconn. Techn. Conf. (IITC), May 24-26, San Francisco, CA, 158 (1999).

[3.45] F. Preston, J. Soc. Glass Technol., **11**, 247 (1927).

[3.46] F. G. Shi, B. Zhao, *Appl. Phys. A*, **67**, 249 (1998).

[3.47] M. Bushnan, R. Rouse, J. E. Leukens, *J. Electrochem. Soc.*, **142(11)**, 3845 (1995).

[3.48] S. R. Runnels, P. Renteln, *Proc. Electrochem. Soc.*, **93-25**, 110 (1993).

[3.49] D, Wang, J. Lee, K. Holland, T. Bibby, S. Beaudoin, T. Cale, *J. Electrochem. Soc.*, **144(3)**, 1121 (1997).

[3.50] J. Tichy, J. A. Levert, L. Shan, S. Danyluk, *J. Electrochem. Soc.*, **146(4)**, 1523 (1999).

[3.51] D. Dornfield, *$17^{th}$ Intl. VLSI Mult. Interconn. Conf (VMIC)*, 105 (2000).

[3.52] J. Luo and D. Dornfield, $18^{th}$ Int. Conf. On Very Large Scale Multilevel Interconnects (VMIC), 281 (2001).

[3.53] J. Luo and D. Dornfield, IEEE Trans. Semiconductor Manufacturing, **14(2)**, 112 (2001).

[3.54] J. Coppeta, C. Rogers, A. Philipossian, F. B. Kaufman, *1st Intl. Conf. on CMP Planar. (CMP-MIC) Conference*, February 13-14, Santa Clara, CA, 307 (1997).

[3.55] S. R. Runnels, M. Eyman, *J. Electrochem. Soc.*, **141(5)**, 1698 (1994).

[3.56] C. Rogers, J. Coppeta, L. Racz, A. Philipossian, F. B. Kaufman, D. Bramono, *J. Elect. Mat.*, **27(10)**, 1082 (1998).

[3.57] S. R. Runnels, *J. Electrochem. Soc.*, **141(7)**, 1900 (1994).

[3.58] S. Sundararajan, D. G. Thakurta, D. W. Schwendeman, S. P. Murarka, W. N. Gill, *J. Electrochem. Soc.*, **146(2)**, (1999).

[3.59] D. G. Thakurta, C. L. Borst, D. W. Schwendeman, R. J. Gutmann, W. N. Gill, *Thin Solid Films*, **366**, 181 (2000).

[3.60] E. Paul, J. Electrochem. Soc., **148**, G355 (2001).

[3.61] E. Paul, J. Electrochem. Soc., **148**, G359 (2001).

[3.62] R.J. Gutmann, D.J. Duquette, P.S. Dutta and W.N. Gill, $6^{th}$ Intl. Conf. On CMP Planar. (CMP-MIC) Conference, 81 (2001).

[3.63] H. S. Fogler, Elements of Chemical Reaction Engineering, $2^{nd}$ Edition, New Jersey: Prentice Hall (1992).

[3.64] C. N. Hinshelwood, The Kinetics of Chemical Change, Oxford: Clarendon Press (1940).

[3.65] A. J. Roosmalen, *Vacuum*, **34(3-4)**, 429 (1984).

[3.66] H. Ishii, Y. Takahashi, *J. Electrochem. Soc,.* **135(6)**, 1539 (1988).

[3.67] S. M. Hsu, R. L. Kabel, *AICHE Jrnl.*, **20(4)**, 713 (1974).

# Chapter 4

# CMP OF BCB AND SILK POLYMERS

Experiments designed and performed to examine the CMP of blanket BCB and SiLK polymer films are presented in this chapter. The films were polished using Cu CMP slurries to simulate the overpolish step that occurs when all metal and liner materials are removed in the damascene patterning process. Two sets of experiments were performed to determine the controlling factors for polymer CMP removal. The first set of experiments explore the effects of various slurries on polymer removal rates, post-CMP surface roughness, and post-CMP surface and interior chemistry. The second set of experiments investigates the effect of reducing the cure time and temperature on polymer CMP removal rates and post-CMP film hardness and modulus. These results are later combined to develop a physically-based explanation and conceptual model of the CMP process applied to low dielectric constant polymers. The experimental procedures and characterization techniques employed are thoroughly presented in Appendix A. This chapter emphasizes CMP results and a mechanistic understanding.

## 4.1 REMOVAL RATE IN COPPER SLURRIES

The first set of experiments used five acidic slurries with alumina abrasives to polish BCB and SiLK polymers under normal cure conditions. Three slurries contain nitric acid with and without surfactants, and two slurries are Rodel QCTT1010-based metal slurries with different size alumina abrasives. The slurry compositions are detailed in Table A3. The recommended cure conditions are 250°C for 30 minutes for BCB and 450°C for 6 minutes for SiLK, as listed in Dow Chemical Company product literature to obtain > 95% film crosslinking during the cure. Polymer removal rate, surface roughness, and chemical alteration are equally important aspects of a viable CMP process, and are investigated as such.

Figures 4.1 and 4.2 illustrate the removal rates obtained for BCB and SiLK polymer films subjected to 2.5 psi, 30 rpm carrier, 30 rpm platen, and 200 ml/min slurry dispensed during CMP. The pad used for these experiments is the SUBA IV soft pad. The nitric acid control slurry with alumina abrasive described in Table A.3 is effective for simplified copper CMP laboratory experiments and was compared with identical slurries containing surfactant (slurries 1 and 2). The control slurry results in unmeasurable removal rates and heavy scratching for both polymers and is not shown in Figure 4.1 and 4.2. Slurries 3 and 4 are commercial slurries (variations of Rodel QCTT1010) with a proprietary chemical composition and alumina abrasives as described in Appendix A.

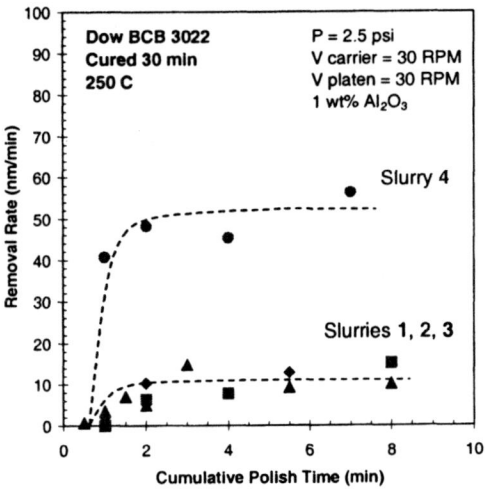

*Figure 4.1.* Removal rate of BCB in slurries 1-4. BCB RR increases with abrasive particle size (from ref [4.1])

The data in Figure 4.1 show that zero removal is obtained during the initial 30-60 seconds of BCB CMP [4.1]. This time period represents a "latency" in CMP removal, which is caused by the hydrophobic nature of the organic BCB polymer. This latency period is attributed to the time required for surfactant additives to adsorb sufficiently on the polymer surface, allowing close contact between the polymer film and the CMP slurry. Surfactant molecules also adsorb on the abrasive particles in the slurry, creating a protective layer between polymer and abrasive due to the hydrophobic/hydrophilic nature of the surfactant groups. All slurries used result in low BCB removal rate, with the largest removal rate approximately 60 nm/min and a typical value of 10 nm/min. Low material removal rate indicates either the lack of a polymer-altering surface reaction, or the formation of a hardened passivating layer on the material surface. However, the elimination of heavy scratching observed with the control slurry with both anionic and nonionic surfactants (presented in Section 4.2) indicates that a hardened, passivating layer is formed.

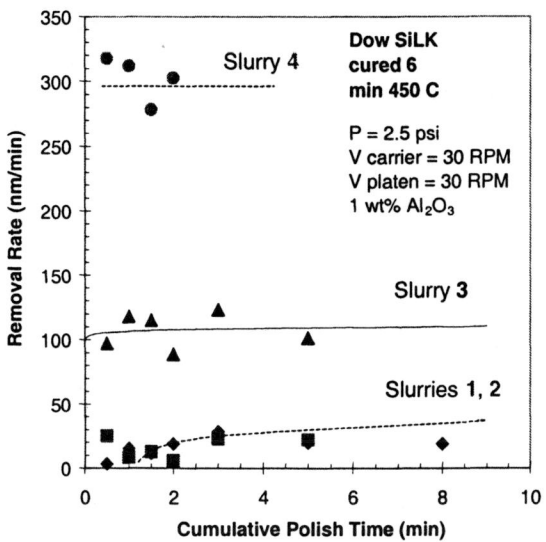

Figure 4.2. Removal rate of SiLK in slurries 1-4. SiLK RR increases with QCTT1010 chemical and abrasive size (from ref [4.1])

During the latency period, slurry is not able to completely wet the polymer surface due to the hydrophobicity of the organic structure. As a consequence, slurry chemicals and abrasive particles do not achieve intimate contact with the polymer surface. At the end of the latency period, once the surfactant has been adsorbed on all available surfaces, abrasive particles and slurry shear combine to remove BCB polymer material at a rate of 10-15 nm/min for 0.05 μm $Al_2O_3$ and 40-55 nm/min for 0.30 μm $Al_2O_3$. The larger abrasive particles result in higher BCB removal rate due to higher shear forces and larger areas of abrasive impact. The latency period observed in BCB with all slurries is evidence that the adsorbed surfactant additive aids in forming an altered polymer surface layer.

Figure 4.2 shows SiLK polymer removal rate data. SiLK CMP exhibits the same latency period as observed in BCB CMP for slurries 1 and 2. Following the surfactant adsorption during this initial 60-90 seconds, slurry wetting of the polymer surface results in chemical and abrasive contact, and low material removal rates of 10-20 nm/min. SiLK exhibits much different behavior when polished with slurries 3 and 4.

In slurries 3 and 4 the oxidizing chemicals in the commercial slurry greatly accelerate the removal of SiLK polymer during CMP. Removal rates are an order of magnitude higher in the presence of the QCTT1010 commercial chemical. This high removal rate indicates a rapid chemical reaction at the SiLK surface, during which structural bonds in the polymer are broken, forming an altered polymer layer. The shear induced during CMP removes the altered SiLK layer, resulting in removal rates that are comparable to those observed for Cu during CMP in these slurries.

In summary, Figures 4.1 and 4.2 show that:
1) QCTT1010 commercial slurry with oxidizing chemistry has little or no effect on the removal rate of BCB polymer.
2) QCTT1010 commercial slurry with oxidizing chemistry substantially increases the removal rate of SiLK polymer.
3) Larger abrasive particles significantly increase the removal rates of both BCB and SiLK polymers.
4) A synergistic effect is present between the mechanical forces and the commercial slurry chemistry during SiLK polymer CMP

In all cases, higher rates and smoother surfaces are measured when surfactant is present in the slurry (based on low, unmeasurable removal rates measured using the control slurry). These removal rates are comparable to BCB and SiLK removal rates of approximately 50 and 450 nm/min respectively measured using undiluted QCTT1010 slurry reported by Küchenmeister, *et al* [4.2].

## 4.2 SURFACE ROUGHNESS

Atomic force microscopy (AFM) scans of BCB and SiLK measure the surface roughness and scratching that results from CMP. High roughness and scratching damage to the film surface suggest largely mechanical abrasion during CMP. Low roughness and scratching suggest the presence of a protective layer on the polymer surface that resists abrasive wear and damage. When used in combination with removal rate and surface analysis, AFM scans can provide qualitative information about the mechanism for polymer removal. Surface scans of area 1.5 µm x 1.5 µm were taken for both BCB and SiLK after CMP with slurries 1 – 4. Representative scans are shown in Figures 4.3, 4.4, 4.5, and 4.6.

Figures 4.3 and 4.5 show the unpolished polymer film surface that is typically observed. Figures 4.4 and 4.6 are representative of the surfaces that remain after CMP of both polymers in slurries 1 – 4. Nanometer-scale scratches can be observed in the post-CMP scans, suggesting material softness. Figure 4.6 also shows a wrinkle defect on the post-CMP SiLK surface. This defect is believed to be the result of inelastic deformation of the SiLK polymer surface under the shear stress developed during CMP [4.3].

*Figure 4.3.* 1.5 µm x 1.5 µm AFM scan of unpolished BCB 5021-32 (from ref [4.1])

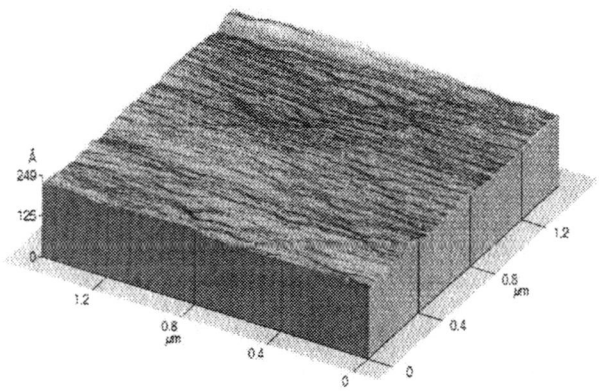

*Figure 4.4.* 1.5 µm x 1.5 µm AFM scan of BCB 5021-32 after 8 minutes of CMP in slurry 3. Removal rate = 10 – 15 nm/min (from ref [4.1])

Table 4.1 contains AFM results for the BCB and SiLK samples following CMP. Prior to CMP the BCB surface is very smooth, and after CMP the BCB roughness increases slightly. CMP with slurries 1 – 4 results in far less roughness and scratching than CMP with the control slurry. Reduced surface damage shows the effect of the surfactant additive on the BCB surface. Adsorbed surfactant protects the surface, aiding in slow, controlled removal and low RMS roughness. The protective action is attributed to

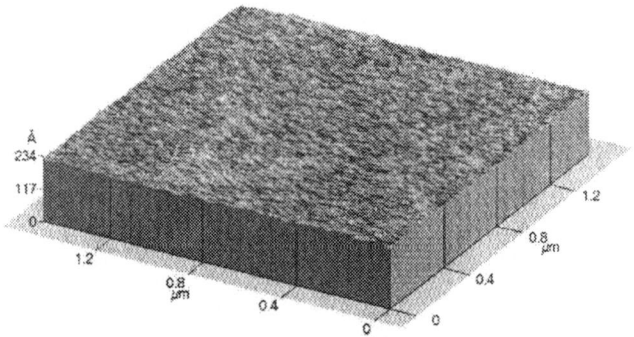

*Figure 4.5.* 1.5 μm x 1.5 μm AFM scan of unpolished SiLK (from ref [4.1])

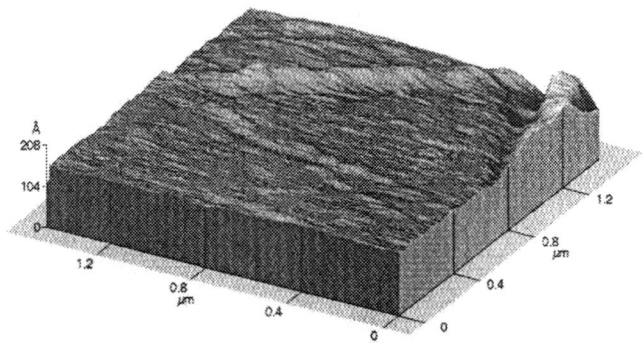

*Figure 4.6.* 1.5 μm x 1.5 μm AFM scan of SiLK after 5 minutes of CMP in slurry 3. Removal rate = 100 – 110 nm/min (from ref [4.1])

surfactant adsorption on the BCB and the slurry particles, providing a buffer between the polymer and the hard $Al_2O_3$. The presence of surfactant also increases removal uniformity and repeatability of thickness measurements after CMP (compared to the control slurry). AFM results for BCB are consistent with observations reported by Yang, *et al.* [4.4].

SiLK polymer shows similar post-CMP roughness results. The control slurry results in low removal rate and high scratching and gouging of the polymer surface. In contrast, smooth surfaces (~ 2.0 nm RMS) are observed following CMP with slurries 1-4. This observation suggests that slurry surfactants protect the SiLK in the same manner that the BCB surface is protected from damage. The fact that the high removal rate slurries (slurries 3 and 4) leave smooth SiLK surfaces suggests that the mechanism for quick SiLK removal does not require scraping and gouging of the SiLK polymer. Rather, the slurry chemistry allows the higher removal rate while the abrasive component scrapes and abrades away the surface with same degree of damage observed for slurries 1 and 2. The slightly higher overall post-CMP RMS roughness of SiLK compared to BCB suggests that SiLK may be softer than BCB, allowing increased roughening of the SiLK surface.

Table 4.1. BCB and SiLK AFM results show smooth post-CMP surfaces[129]

| | Slurry | Time (min) | Removal Rate (nm/min) | RMS Roughness (nm) | Comments |
|---|---|---|---|---|---|
| BCB | Unpol | --- | --- | 0.45 | smooth, as deposited |
| BCB | Control | 8.0 | ~ 0 | not measured | extensive visible scratching |
| BCB | Slurry 1 | 5.5 | 10 – 15 | 1.4 | Controlled abrasion after latency period |
| BCB | Slurry 2 | 8.0 | 10 – 15 | 0.7 | Controlled abrasion after latency period |
| BCB | Slurry 3 | 8.0 | 10 – 15 | 0.5 | Controlled abrasion after latency period |
| BCB | Slurry 4 | 7.0 | 45 – 55 | 1.3 | Controlled abrasion after latency period |
| SiLK | Unpol | --- | --- | 0.45 | smooth, as deposited |
| SiLK | Control | 8.0 | ~ 0 | not measured | extensive visible scratching |
| SiLK | Slurry 1 | 8.0 | 20 | 1.6 | Controlled abrasion after latency period |
| SiLK | Slurry 2 | 5.0 | 20 | 1.1 | Controlled abrasion after latency period |
| SiLK | Slurry 3 | 5.0 | 110 | 0.8 | Synergistic chemical/mechanical CMP |
| SiLK | Slurry 4 | 2.0 | 300 | 2.0 | Synergistic chemical/mechanical CMP |

Comparison of the AFM results from the nitric acid and QCTT1010 slurries provides additional information about the mechanism for rapid SiLK removal observed with slurries 3 and 4. The high removal rate and smooth post-CMP surface of SiLK is the result of a synergistic altered-layer reaction. This result is in contrast to high-scratching, non-uniform removal of SiLK observed with the control slurry. "Synergistic chemical/mechanical CMP" means that the slurry chemicals and abrasive act together in the CMP process. The slurry chemicals react with the polymer surface, break structural bonds, and produce a weakened surface layer. Abrasive particles and slurry shear remove the weakened layer and expose an unreacted polymer surface. The surface reaction/removal process repeats, leading to rapid polymer removal but little physical damage to the polymer surface.

## 4.3 SURFACE AND BULK FILM CHEMISTRY

X-ray photoelectron spectroscopy (XPS) is used to measure change in polymer surface atomic concentrations as a result of CMP. After measuring removal rates in all slurries, slurry 4 was selected for the XPS chemistry experiments. This choice was made due to the low removal rate observed when polishing BCB and the high removal rate observed when polishing SiLK. The hope was to gain some insight into the mechanism that would allow the same slurry to polish the different polymer materials in such different ways.

### 4.3.1 Angle-Resolved Surface Results

Angle-resolved XPS was performed on BCB and SiLK samples polished with slurry 4. The angle-resolved technique probes 1 to 10 nm deep by varying the angle of incident x-rays. Shallower incident angles take information from the region closer to the surface. Specifically, 20, 45, and 90 degree angles give composition and bonding information for 1-2, 5, and 10 nm depths, providing a surface profile.

A survey spectrum for unpolished BCB is shown in Figure 4.7. The main peaks in the spectrum are carbon, suggesting that the polymer is largely carbonaceous, silicon, and oxygen that exist in the form of the -Si-O-Si- linkage in the BCB structure. Results for BCB after CMP in slurry 4 are shown in Figures 4.8, 4.9, and 4.10. Figure 4.8 is a survey spectrum that shows the entire range of binding energies, and includes peaks for C, O, Si, and H at each of the measurement angles. Figures 4.9 and 4.10 are expansions of the binding energy axis that are used to isolate the changes in the oxygen and carbon bonding observed after CMP.

The survey spectrum shows that after CMP, the BCB composition is relatively constant throughout the first 10 nm of the film surface. Comparison to the unpolished BCB spectrum shows that the oxygen content has noticeably increased following CMP in slurry 4. The oxygen binding energy window shows a change in the oxygen bonding in the BCB surface after CMP.

Prior to CMP, the O peak is dominant at 532.5 eV, the energy of -Si-O- bonding. A smaller peak is observed at the left shoulder near 534 eV, the energy of -C-O- bonding. After CMP, the two peaks are of the same intensity, resulting in one broad peak. This is evidence that the carbon at the surface of the BCB sample is oxidized during CMP. The carbon binding energy window also shows evidence of carbon oxidation. Initially, the

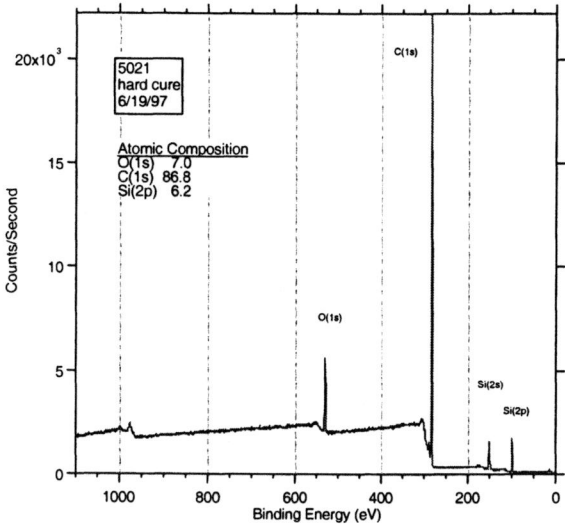

*Figure 4.7.* XPS survey spectrum for unpolished BCB 5021

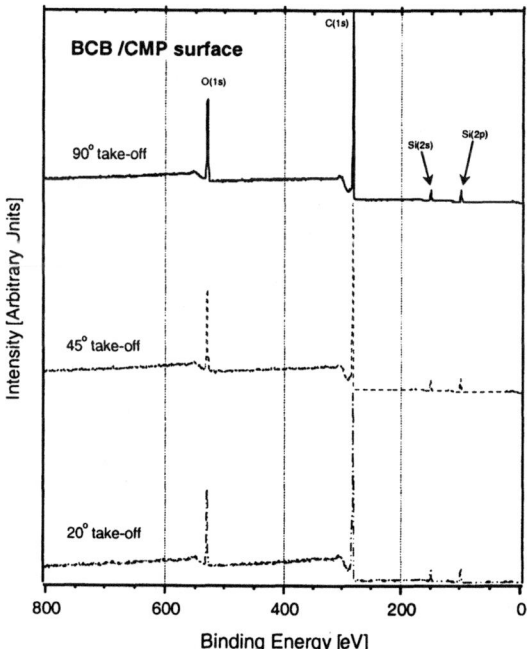

*Figure 4.8.* Survey spectrum for angle resolved XPS measurements of BCB after CMP in slurry 4

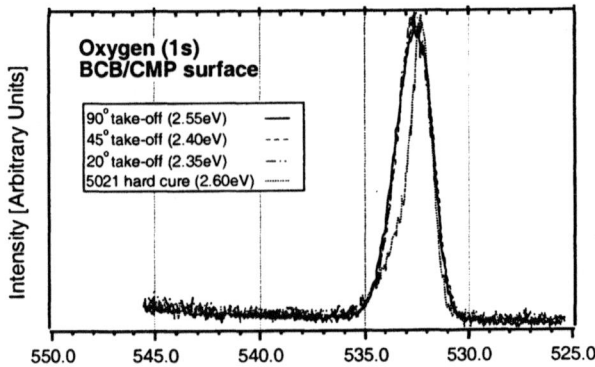

*Figure 4.9.* Oxygen binding energy window for BCB after CMP in slurry 4

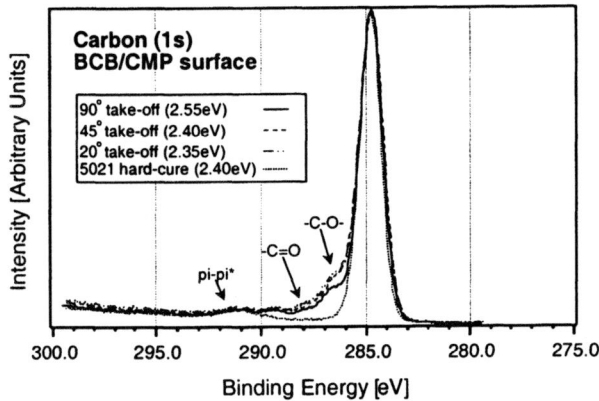

*Figure 4.10.* Carbon binding energy window for BCB after CMP in slurry 4

carbon peak is a single sharp peak at 285 eV, the energy of -C-C- bonding. After CMP, shouldering peaks become readily apparent at energies of 286 and 287 eV. These peaks represent an increase in the -C-O- and -C=O bonding at the BCB surface.

A survey spectrum for unpolished SiLK is shown in Figure 4.11. The main peak in the spectrum is carbon, showing that the polymer is almost completely carbonaceous. A small oxygen peak exists, showing that there is a low percentage of oxygen in the cured film. Figure 4.12 is a survey spectrum that shows the entire range of SiLK binding energies, and includes peaks for C, O, and H at each of the measurement angles after CMP. Figures 4.13 and 4.14 are expansions of the binding energy axis that are used to isolate the changes in the oxygen and carbon bonding observed after CMP.

The post-CMP survey spectrum shows that after CMP the SiLK composition shows a gradient in oxygen throughout the first 10 nm of the film surface. Comparison to the unpolished SiLK spectrum shows that the oxygen content has noticeably increased following CMP in slurry 4. The oxygen binding energy window shows a change in the oxygen bonding in the SiLK surface after CMP. Prior to CMP the O peak is dominant at 534 eV, the energy of -C-O- bonding. A shouldering peak is observed at 523 eV, most likely due to free oxygen at the sample surface. After CMP the oxygen peak greatly increases in intensity, and narrows to a single energy. This is evidence that the carbon at the surface of the SiLK sample is also oxidized during CMP. However, this surface reaction enables the high removal rates observed in Figure 4.2. The carbon binding energy window also shows evidence of carbon oxidation. Initially, the carbon peak is a single sharp peak at 285 eV, the energy of -C-C- bonding. After CMP, shouldering peaks become apparent at energies of 286 and 287 eV. These peaks represent an significant increase in the -C-O- and -C=O bonding at the SiLK surface due to CMP in slurry 4.

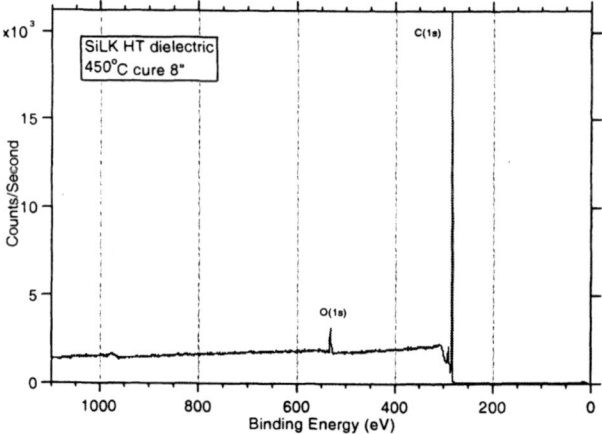

*Figure 4.11* XPS survey spectrum for unpolished SiLK

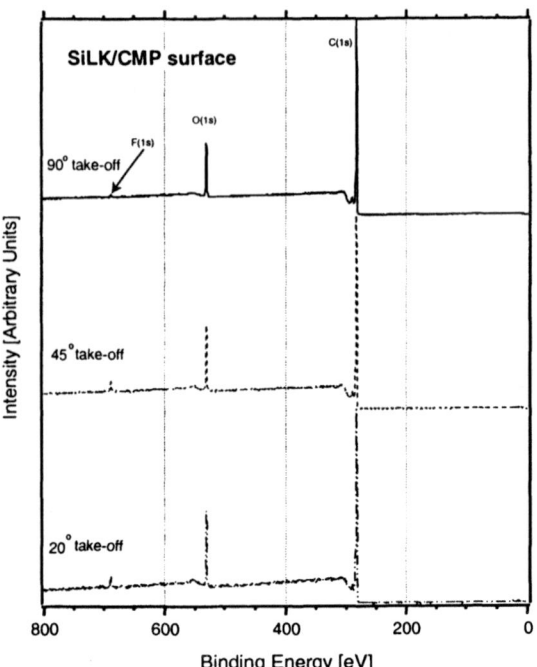

*Figure 4.12.* Survey spectrum for angle resolved XPS measurements of SiLK after CMP in slurry 4

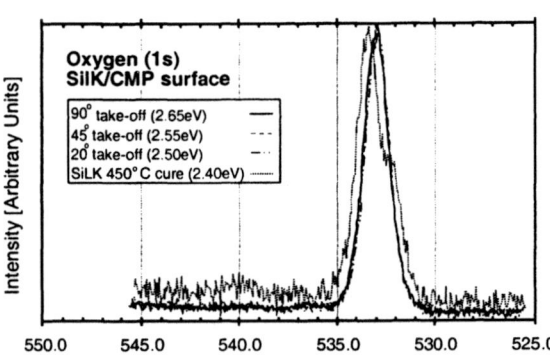

*Figure 4.13* Oxygen binding energy window for SiLK after CMP in slurry 4

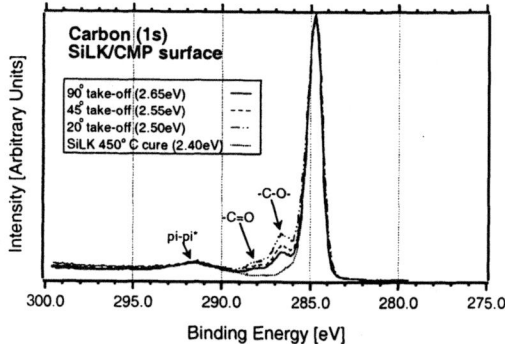

*Figure 4.14.* Carbon binding energy window for SiLK after CMP in slurry 4

Further evidence of the formation of an altered surface layer in both BCB and SiLK materials is shown in Table 4.2 in the form of atomic concentrations in the polymer surfaces. As-deposited BCB and SiLK contain bulk oxygen concentrations of 3 at% and ~ 0 at% respectively. Following CMP, the BCB surface oxygen concentration increases to 14 at% (consistent with previously reported results [4.4]). The SiLK surface oxygen concentration increases to 10 at%. The increase in the surface oxygen for both polymers is evidence of an altered surface layer greater than 10 nm.

Constant oxygen concentration (14 at%) in the top 10 nm of the BCB surface suggests that oxidation of the polymer proceeds to equilibrium due to extended contact with the slurry. Because of slow removal, the same BCB material remains at the surface longer, allowing oxidation to reach equilibrium. Oxidation may stabilize or harden the BCB surface, resulting in resistance to shear and abrasion.

*Table 4.2.* XPS surface atomic concentrations for BCB and SiLK following CMP in slurry 4

|  | Angle (deg) | Depth (nm) | F (1s) (atom %) | O (1s) (atom %) | C (1s) (atom %) | Si (2p) (atom %) |
|---|---|---|---|---|---|---|
| BCB Unpol | 90 | 10 | - - | 3.5 (±2.2) | 90.5 (±4.0) | 6.0 (±0.4) |
| BCB | 20 | 1-2 | - - | 14.3 (±2.3) | 80.8 (±4.0) | 5.0 (±1.2) |
| BCB | 45 | 5 | - - | 14.3 (±0.7) | 81.2 (±4.1) | 4.5 (±0.2) |
| BCB | 90 | 10 | - - | 14.8 (±2.2) | 80.6 (±2.0) | 4.6 (±0.4) |
| SiLK Unpol | 90 | 10 | - - | 0.5 (±0.4) | 99.5 (±0.8) | - - |
| SiLK | 20 | 1-2 | 0.2 (±0.1) | 12.8 (±0.6) | 87.2 (±4.4) | - - |
| SiLK | 45 | 5 | 0.2 (±0.1) | 11.1 (±0.6) | 88.9 (±4.4) | - - |
| SiLK | 90 | 10 | 0.3 (±0.1) | 10.0 (±0.4) | 89.8 (±0.8) | - - |

The SiLK contains an oxygen gradient from 13 at% O in the first 1-2 nm to 10 at% O at a depth of 10 nm. During CMP with QCTT1010 slurry, the slurry chemistry appears to break structural bonds in the SiLK surface. This reaction proceeds until the polymer surface has been weakened sufficiently to allow material removal, exposing the less-oxidized underlying polymer structure. Repetition of this process results in accelerated SiLK removal.

### 4.3.2 Depth Profiling Results

The depth profiling XPS technique measures atomic concentration and bonding character at different depths into the sample interior as a low-energy ion sputters away the film. Measurements are taken each fifth sputter cycle over a total of 25 cycles to provide chemical information beginning at the post-CMP polymer surface and penetrating approximately 100 nm deep into the polymer samples. This depth is an order of magnitude deeper than the angle resolved technique. Depth profiling XPS allows a comparison of the surface and bulk polymer composition, to determine the presence of an altered polymer surface layer and observe any interior chemical alteration.

XPS depth profiles were measured for both polymers after CMP in slurries 1 and 4 to compare the very low pH nitric acid slurry to the QCTT1010. Spectrum are included in Appendix B, in Figures B-1, B-2, and B-3 for BCB and Figures B-4, B-5, and B-6 for SiLK. The composition information measured in the initial XPS cycles correspond well with the surface XPS data discussed previously. Both BCB and SiLK polymers were found to increase in oxygen surface concentration following CMP with slurries 1 and 4, as tabulated in Table 4.3.

The BCB surface oxygen content increases from 3.5 at% as-deposited to 8.5 and 10.1 at% after CMP with slurries 1 and 4 respectively. The SiLK oxygen content increases from 0.5 at% as-deposited to 5.5 and 7.3 at% after CMP. However, the measurements show negligible chemical alteration of the polymer interior after CMP. The data for both BCB and SiLK show that the interior atomic composition remains constant at the expected concentrations of O, C, and Si. The low atomic percent (near the error level of the technique) nitrogen observed following CMP with slurry 4 is believed to be from an $NH_3$ component in the QCTT1010 slurry. The presence of aluminum at the surface of the SiLK sample following CMP with slurries 1 and 4 is attributed to $Al_2O_3$ abrasive particles adhering to the polymer surface within the 800 μm x 800 μm measurement area. Several $Al_2O_3$ particles were observed in large-area AFM scans, since the samples were not brush cleaned following CMP. In accordance with the aluminum artefact, a stoichiometric amount of oxygen ratioed to $Al_2O_3$ should also be subtracted from the data. Such a subtraction yields 0.55 at% oxygen in the SiLK bulk following CMP in slurry 4, a negligible bulk material increase.

Depth XPS supports the hypothesis that oxidation due to slurry chemistry results in an altered surface layer that allows SiLK polymer removal and passivates the BCB surface, while leaving the interior polymer material unaffected.

Table 4.3. Surface and interior compositions of BCB and SiLK from depth profiling XPS.

| | Slurry | Region | Al (2p)* (atom %) | O (1s) (atom %) | C (1s) (atom %) | Si (2p) (atom %) | N (1s)* (atom %) |
|---|---|---|---|---|---|---|---|
| BCB | Unpol | Surface | -- | 3.5 | 90.5 | 6.0 | -- |
| BCB | Slurry 1 | Surface | 3.4 | 8.5 | 82.5 | 5.1 | -- |
| BCB | Slurry 4 | Surface | -- | 10.1 | 85.1 | 4.8 | -- |
| BCB | Unpol | Bulk | -- | 2.5 | 91.0 | 6.5 | -- |
| BCB | Slurry 1 | Bulk | -- | 3.5 | 90.0 | 6.5 | -- |
| BCB | Slurry 4 | Bulk | -- | 3.1 | 89.2 | 6.0 | 1.7 |
| SiLK | Unpol | Surface | -- | 0.5 | 99.5 | -- | -- |
| SiLK | Slurry 1 | Surface | -- | 5.5 | 94.5 | -- | -- |
| SiLK | Slurry 4 | Surface | 0.9 | 7.3 | 91.8 | -- | -- |
| SiLK | Unpol | Bulk | -- | 0.5 | 99.5 | -- | -- |
| SiLK | Slurry 1 | Bulk | -- | 0.5 | 99.5 | -- | -- |
| SiLK | Slurry 4 | Bulk | 0.75 | 1.55 | 95.9 | -- | 1.8 |

*- Aluminum and Nitrogen impurities are from alumina abrasive and ammonia present in slurries 1 and 4. Improved surface cleaning should eliminate these impurities.

## 4.4 EFFECT OF CURE CONDITIONS ON BCB AND SILK REMOVAL

The second set of experiments was performed to examine the chemical-mechanical planarization of BCB and SiLK cured at different temperatures for different periods of time. Slurries 1, 3, and 4 listed in Table A.3 were used, with abrasive particle size again varied from 0.05 μm to 0.30 μm. These slurries were chosen to compare the effects of the low removal rate surfactant slurry to the high removal rate QCTT1010 slurry, and to examine the effect of the abrasive size in the QCTT1010 slurry. This section shows the effects of both the temperature at which the polymer is cured and the time allowed for curing. The effect of curing was measured through material removal rates and pre- and post- CMP measurement of mechanical film properties.

### 4.4.1 Variation in Cure Conditions

Table A.1 lists the BCB and SiLK films that were used in the study. The BCB cure cycle was increased in both temperature and duration from its recommended 30 minutes at 250°C. The SiLK cure cycle was decreased in temperature from its recommended 450°C.

Less aggressive thermal processing may allow manufacturers to fabricate the interconnect levels at temperatures ranging from 350 to 400°C. For this reason, the properties of SiLK that is cured at a lower temperature of 400°C is of interest. BCB has more margin between the prescribed cure temperature and the interconnect thermal budget. For this reason, a BCB that is cured at slightly higher temperature (although still substantially below the thermal budget) might exhibit beneficial properties for CMP integration.

### 4.4.2 Effect of Cure Conditions on Removal Rate

The previous work with acidic copper slurries shows removal rates of 10 – 50 nm/min for BCB cured at 250°C and 20 – 300 nm/min for SiLK cured at 450°C. BCB cure temperature was increased to 300°C and SiLK cure temperature was decreased to 400°C to examine how the cure temperature effects the CMP removal rate. Lower cure temperature allows less complete crosslinking of the polymer films, since the reduced thermal energy in the polymer reactive groups allows less chain flexibility and motion. The result is that fewer reactive sites are able to locate one another, react, and crosslink. Differences in the cure cycles for BCB and SiLK were examined using slurries 2 and 3, the Rodel QCTT1010 slurries. The data in Figures 4.15 and 4.16 show a trend between removal rate and cure temperature for both BCB and SiLK materials.

Figure 4.15 shows that CMP removal rates for BCB 3022 and 5021 increase as both cure time and cure temperature decrease, but the latency period for CMP removal is unaffected [4.5]. The lowest removal rate of 2 nm/min results from long cure duration (120 min) and high temperature (300°C) for BCB. Although increased removal rate is observed at the lower cure temperature (250°C), the rate of 50 nm/min remains low. The low removal rate of BCB in all slurries is attributed to a slow mechanically dominated polish. The data suggest that BCB cured at a lower cure temperature retains its chemical stability, but loses some of its mechanical strength due to less complete crosslinking.

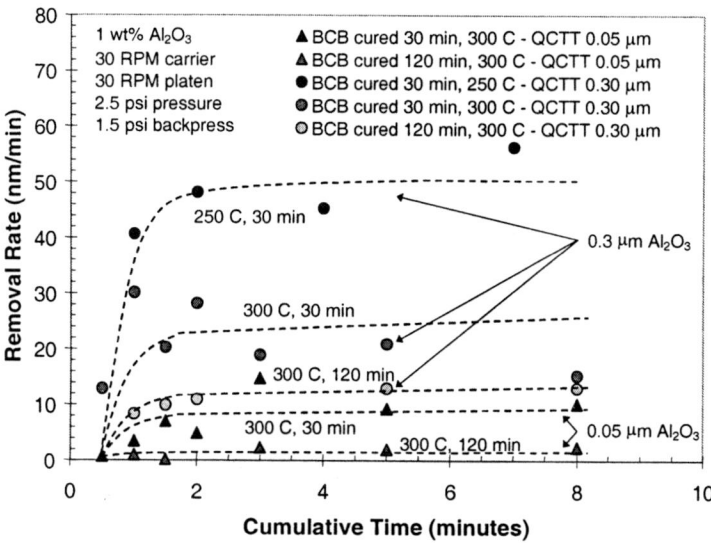

*Figure 4.15.* CMP removal rates in slurries 3 and 4 for BCB 3022 and 5021, cured at different temperatures and times. Longer duration and higher cure temperature results in lower removal rate in both slurries (from ref [4.5])

Figure 4.16 shows that CMP removal rates for SiLK also increase as cure temperature decreases. A latency period for removal is not observed within the first 30 seconds of CMP, although it is possible that the latency exists on a much smaller time scale not captured by these experiments. The lowest removal rate of 150 nm/min results from high cure temperature and smaller slurry abrasive particles. This removal rate is still an order of magnitude higher than the removal rate of BCB in this slurry, suggesting that slurry chemistry works together with shear and abrasion to remove the SiLK material. The highest removal rate of 830 nm/min occurs for mild cure temperature (400°C) and larger abrasive size. The high removal rates in slurries 2 and 3 are attributed to a reactive slurry chemistry that alters the SiLK surface, leaving a softer, less crosslinked surface layer that is rapidly removed by shear and abrasion.

The chemical-mechanical synergistic action of CMP is more effective when the SiLK film is cured at a lower temperature. Lower temperature and less complete crosslinking have a twofold effect on the removal rate. First, reduced crosslinking results in reduced film strength, allowing an increase in the mechanical contribution to removal. Second, reduced crosslinking may leave greater accessibility for slurry chemistry to attack the SiLK film,

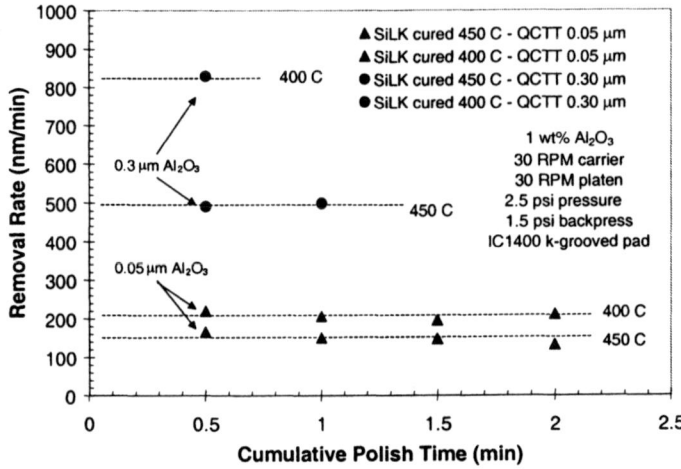

*Figure 4.16.* CMP removal rates in slurries 3 and 4 for SiLK cured at different temperatures. Higher temperature cure results in lower removal rate in both slurries (from ref [4.5])

thus enhancing the slurry chemical contribution to CMP. The reduced cure temperature results in an increase in removal rate by a factor of ~ 1.5 compared to conventionally cured SiLK. A difference in removal rate is also observed when SiLK is polished with the two different Rodel pads, SUBA IV and IC 1400 k-grooved. Comparison of Figures 4.2 and 4.16 show that using the IC1400 k-grooved pad in place of SUBA IV results in a 50% increase in removal rate for SiLK cured at 450°C. Previous work dealing with different pad materials suggests that pad hardness and ability for slurry transport have significant effects on CMP removal rate of metals and oxides [4.6-4.8]. Our results support this claim in the specific case of SiLK, where the IC 1400 pad may have higher removal rate due to increased hardness and better ability to transport slurry to and from the polymer surface and site of chemical attack.

## 4.5 EFFECT OF CMP ON BCB AND SILK FILM HARDNESS

Nanoindentation measurements were taken to compare and contrast polished and unpolished BCB and SiLK with different cure conditions. Load/depth data is used to calculate the hardness and elastic modulus of the thin films. This information has been combined with removal rate and surface topography measurements to produce information about polymer structural integrity as a result of the CMP process.

An example plot of a polymer nanoindentation load/depth curve is shown in Figure 4.17. SiLK films cured at two different temperatures were measured and compared to a BCB film. The hardness for the unpolished samples follows the order BCB (1.008 GPa) > SiLK cured at 450°C (0.774 GPa) > SiLK cured at 400°C (0.748 GPa). The silicon in the BCB backbone results in a harder polymer matrix, while the higher cure temperature for the SiLK results in more complete crosslinking than the lower cure temperature. The difference in hardness of the 400 and 450°C cured samples is small, but is consistently measured across five separate sample points.

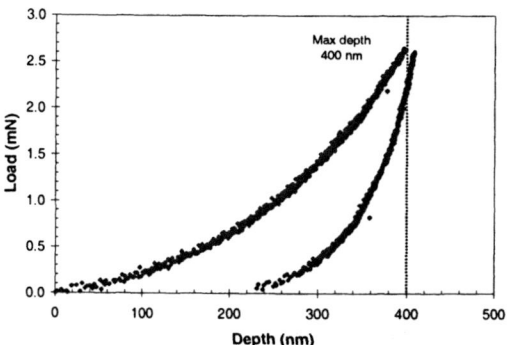

*Figure 4.17.* Load vs. depth curve for nanoindentation of unpolished SiLK cured at 450 °C, measured to 400 nm maximum depth. Hardness and elastic modulus are calculated from this curve. From ref [4.9].

Nanoindentation measurements taken following CMP provide insight into the polishing process for SiLK with slurries 1 and 3. From previous removal and topography results, SiLK CMP with slurry 1 is characterized as a slow, mechanically dominated process. The nanoindentation measurements in Table 4.4 suggest that mechanical abrasion in slurry 1 causes a reduction in the hardness and elastic modulus of both SiLK samples. The hardness measured by nanoindentation is related to the failure of the material under load, and depends on the defectivity of the film [4.6].

Mechanically-dominated CMP in slurry 1 makes the SiLK less hard, possibly due to stress-induced defects in the polymer as a result of the abrasion of the surface material. The elastic modulus measured by nanoindentation is related to the inability of the material to respond elastically when a load is removed. A reduced modulus suggests that the SiLK becomes less elastic as a result of the mechanical forces imparted during CMP in slurry 1. Figure 4.18 shows a schematic diagram of how film defectivity imparted by CMP can be reflected in the microhardness measurement.

Table 4.4. BCB and SiLK nanoindentation results before and after CMP in slurries 1 and 3.

| | Cure Conditions | CMP Conditions | Sample Thickness (nm) | Nanoindentation Hardness (GPa) | Nanoindentation Modulus (GPa) |
|---|---|---|---|---|---|
| BCB | 300°C, 120 min | Unpolished | 1200 | 1.008 | 22.8 |
| SiLK | 400°C, 6 min | Unpolished | 1200 | 0.748 | 20.58 |
| SiLK | 450°C, 6 min | Unpolished | 1200 | 0.774 | 18.42 |
| SiLK | 400°C, 6 min | Slurry 1, 4 min | 990 | 0.510 | 4.64 |
| SiLK | 450°C, 6 min | Slurry 1, 4 min | 980 | 0.608 | 13.74 |
| SiLK | 400°C, 6 min | Slurry 3, 2 min | 730 | 0.768 | 34.84 |
| SiLK | 450°C, 6 min | Slurry 3, 2 min | 640 | 1.048 | 24.3 |

In contrast, CMP in slurry 3 does not detrimentally effect the SiLK hardness or elastic modulus. Previous results suggest that slurry 3 attacks the SiLK film with combined chemical-mechanical action, resulting in higher removal rate by continuously removing a rapidly forming altered surface layer. The concept of rapid film removal resulting in less film damage is supported by the data in Table 4.4, where hardness and modulus are not reduced following CMP. Silicon substrate effects might effect these measurements due to the reduced thickness of the SiLK films after polishing and the relatively thick underlying silicon substrate. Since CMP in slurry 3 removes the films rapidly, the thickness of the measured films is only two or three times the measurement indentation depth. Higher hardness and elastic modulus values for these thinner films may be a reflection of the underlying Si substrate properties.

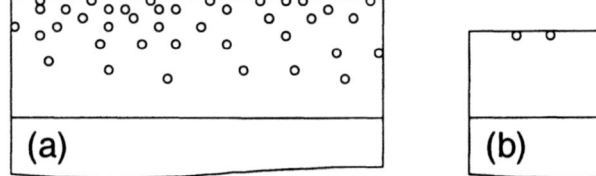

Figure 4.18. Schematic diagrams of (a) BCB film following CMP in slurry 3 (low removal rate, high stress/abrasion resulting in film defectivity) and (b) SiLK film following CMP in slurry 3 (high removal rate, continuous film removal minimizes propagation of stress through film and minimizes defectivity).

## 4.6 COMPARISON OF BCB AND SILK CMP WITH OTHER POLYMER CMP

The CMP of BCB and SiLK offers insights into the CMP of other low-κ polymers and dielectrics. The relative success in polishing these two polymers, as described in sections 4.1-4.5, is largely a result of their chemical structure and reactivity. As mentioned in Chapter 1, the classical model for CMP involves an equilibrium between chemical reaction and physical abrasion of the material to be polished. The ability to obtain a removal rate on the order of hundreds of nanometers per minute without damaging the film physically or electrically relies heavily on this balance between chemical alteration and physical removal.

The CMP of BCB and SiLK is enabled by a combination of material physical strength and a limited chemical reactivity. BCB and SiLK are developed from partially crosslinked monomer (oligomeric) solutions to form a densely crosslinked post-cure matrix that provides physical and thermal stability that is much improved over many polymers. The dense crosslinking is a result of each monomer having several reactive sites which can each become linked to other monomers. For BCB, each monomer has four sties, for SiLK, each monomer has four to eight (see Table 2.7). Thermoplastics such as polyimides, parylenes, and most poly(arylene) ethers form longer chains with two terminal links, and thus do no achieve as high a degree of crosslinking. With higher crosslinking comes a stronger physical structure that is more likely to withstand the pressures, friction, and abrasion developed during the CMP process. BCB and SiLK owe their CMP compatibility in part to a strong, crosslinked structure.

The limited reactivity of BCB and SiLK also derives from their chemical structure. Polymers are developed in many applications to have a high resistance to chemical attack. This is a necessary result of their lower bond polarizability and the aromatic stability of benzene rings in the polymer backbone. Polymer chains have, however, exhibited the tendency to swell in the presence of surfactants, or "swelling agents" [4.10-4.11]. Surfactants are very useful in any system that contains aqueous chemistries (such as DI-water based CMP slurries) and insoluble materials (such as polymers) [4.11]. BCB has exhibited the tendency to interact with surfactant agents to allow CMP removal with low scratching and surface damage. One hypothesis is that the benzene rings in the BCB monomer unit interact with surfactants such as Triton-X 100 [4.12] (See Figure 4.19) and DowFAX [4.13] varieties (Figure 4.20) in an electron-sharing interaction that allows closer contact with the aqueous slurry and possibly enables slow chemical reaction [4.14]. The combination of slight swelling, enhanced chemical reactivity, and a "cushioned" coating of surfactant at the BCB surface result in an "altered layer" for CMP removal that is not exhibited in other polymers.

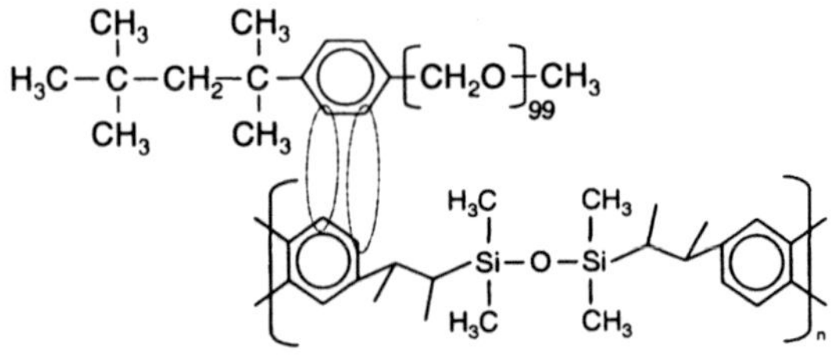

*Figure 4.19* Hypothesized aromatic electron interaction between Triton-X 100 surfactant and BCB monomer unit [4.14].

The chemical reactivity of SiLK is more enhanced than the reactivity of BCB, although the specific reaction, on the basis of organic chemistry, is not well understood. The CMP of SiLK involves a chemical reaction contribution different than the classical model for CMP of more reactive metals such as copper and tungsten. These more reactive metals have well documented oxidation potentials and electrochemical reactions to form the oxidized or reduced product that may be removed by the stress and abrasion developed during CMP. SiLK, however, is a densely crosslinked, almost entirely aromatic structure consisting of electron-resonance-stabilized carbon bonds which are far less susceptible to chemical attack. All the same, the simple slurries described in Appendix A are quite effective at breaking structural bonds in the SiLK structure, enabling high removal rates and low scratching. Other polymers discussed in the literature are polished in a much more abrasive and damaging manner [4.15-4.17].

*Figure 4.20* Hypothesized aromatic electron interaction between DowFAX surfactant and BCB monomer unit.

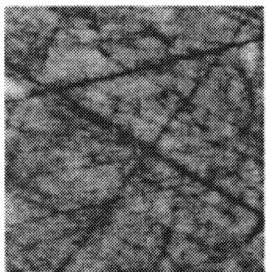

*Figure 4.21* AFM image of scratched and gouged polymer post-CMP

Thus the key to the successful polishing of BCB and SiLK in a controlled, low damage manner lies in the combination of their physical strength and moderate chemical reactivity to slurry wetting or reactive agents. These properties may enable BCB or SiLK to be used in a Cu damascene integration scheme without the necessity of a higher-κ capping layer such as TEOS (which increases the effective film dielectric constant).

## 4.7 SUMMARY

Removal rate, surface topography, and chemical composition measurements show the effects of copper slurries on BCB and SiLK polymers. These measurements show the physical and chemical changes that are induced in the polymers by exposure to slurry chemicals and abrasive. This information is valuable when considering polymer ILDs for reduced capacitance and improved performance in copper/low-κ damascene interconnects.

BCB has a low CMP removal rate in the copper slurries studied. However, reduced BCB cure time and temperature result in slightly higher CMP removal rates due to less complete polymer crosslinking (less structural stability). During CMP, slurries containing surfactant effectively wet the BCB surface, allowing nitric acid and commercial QCTT1010 chemistries to oxidize the BCB surface during CMP. Following CMP, the BCB surface shows an increase in oxygen content, suggesting the formation of a passivating oxidized layer. The bulk BCB material shows no chemical change. Low removal rate and low surface roughness following CMP suggest that the oxidation reaction passivates the BCB surface. Initially, surfactant adsorbs on the polymer surface, enhancing surface wetting and creating a protective layer. The slurry chemistry does not break structural bonds but may oxidize the surface to form a protective layer. The combined protection of oxidation and surfactant adsorption inhibits CMP removal and surface scratching.

SiLK has a low removal rate and low surface scratching in HNO$_3$/surfactant slurries. SiLK removal rate shows an increased dependence on cure temperature, with lower cure temperature resulting in higher removal rate due to two factors: less complete polymer crosslinking resulting in (1) less structural stability and (2) more sites available for attack by the slurry chemistry. A large increase in removal rate is observed when polishing SiLK in QCTT1010 slurry. Although the removal rate is increased by an order of magnitude, surface roughness does not increase greatly. The increased SiLK removal rate is attributed to the synergism of chemical alteration and mechanical abrasion during CMP. Initially, surfactant adsorbs on the polymer surface, enhancing surface wetting and creating a protective layer. Slurry chemistry reacts with the polymer surface, breaking structural polymer bonds and forming a weakened surface layer. Shear stress and abrasion remove the altered layer, resulting in high removal rate. Thus the reactivity of the slurry with polymer structural bonds determines whether polymer CMP removal is slow or rapid.

Hardness measurements following CMP suggest that a CMP process with low removal rate (such as nitric acid slurries with BCB and SiLK) may cause structural defects that result in reduced film hardness. However, a CMP process with high removal rate due to an altered layer reaction (such as QCTT1010 with SiLK) does not result in reduced film hardness.

A fundamentally-based model for polymer CMP is presented in Chapter 6 based on the removal rate, surface topography, and chemical alteration results presented in this chapter. The model includes chemical and surfactant diffusion, surfactant adsorption, and material removal in a first principles-based low-$\kappa$ CMP model.

## 4.8 REFERENCES

[4.1] C. L. Borst, D. G. Thakurta, R. J. Gutmann, W. N. Gill, *J. Electrochem. Soc.*, **146**(11), 4309 (1999).
[4.2] F. Küchenmeister, Z. Stavreva, U. Schubert, K. Richter, C. Wenzel, *Adv. Metal. Conf.*, Colorado Springs, CO, (1998).
[4.3] D. Towery and M. Fury, *J. Elect. Mat.*, **27**(10), 1088 (1998).
[4.4] G.-R. Yang, Y.-P. Zhao, J. M. Neirynck, S. P. Murarka, R. J. Gutmann, *J. Electrochem. Soc.*, **144**(9), 3249 (1997).
[4.5] C. L. Borst, W. N. Gill, and R. J. Gutmann, *Int. Jrnl. of Chemical-Mechanical Planarization for On-Chip Interconnection*, **1**(1), 26 (2000).
[4.6] F. G. Shi, B. Zhao, *Appl. Phys. A*, **67**, 249 (1998).
[4.7] C. Rogers, J. Coppeta, L. Racz, A. Philipossian, F. B. Kaufman, D. Bramono, *J. Elect. Mat.* **27**(10), 1082 (1998).
[4.8] D. G. Thakurta, C. L. Borst, D. W. Schwendeman, R. J. Gutmann, W. N. Gill, *Thin Solid Films*, **366**, 181 (2000).

[4.9]   C. L. Borst, W. N. Gill, and R. J. Gutmann, *Int. Jrnl. of Chemical-Mechanical Planarization for On-Chip Interconnection*, **1(1)**, 26 (2000).
[4.10]  T. F. Thadros (ed.), Surfactants, Academic Press, 1984.
[4.11]  C. L. Borst, Chemical-Mechanical Planarization of Low-Dielectric Constant Polymers in Copper Slurries, M.S. Thesis, Rensselaer Polytechnic Institute, Troy, NY (1999).
[4.12]  http://www.sigma-aldrich.com/sigma/proddata/t6878x.htm
[4.13]  http://www.dow.com/dowfax/clean/prod.html
[4.14]  J. M. Neirynck, Cu/Polymer Damascene Interconnects: Elimination of High-Resistivity Metallic Liners, Ph.D. Thesis, Rensselaer Polytechnic Institute, Troy, NY (1998).
[4.15]  D. Towery and M. Fury, *J. Elect. Mat.*, **27(10)**, 1088 (1998).
[4.16]  J. M. Neirynck, G.-R. Yang, S. P. Murarka, R. J. Gutmann, *Thin Solid Films* **290**, 447 (1996).
[4.17]  D. Permana, S. P. Murarka, M. G. Lee, S. I. Beilin in: R. Havemann, J. Schmitz, H. Komiyama, K. Tsubouchi, *Advanced Metallization and Interconnect Systems for ULSI Applications in 1996*, Boston, USA, October 1-3, 1996, Proceedings of Advanced Metallization and Interconnect Systems for ULSI Applications in 1996, 539 (1997).

# Chapter 5

# CMP OF ORGANOSILICATE GLASSES

The previous chapter outlined experiments that were used to characterize the CMP of low-κ polymer films. In this chapter, we develop a similar understanding for organosilicate (OSG) films, utilizing the low-κ polymer results and the well established understanding of silicon dioxide CMP.

The OSG films studied here vary in carbon content and density, allowing a comparison of film properties and the development of mechanisms for observed differences in CMP performance. The films have been polished with various slurries that are used for CMP to represent the overpolish process in copper damascene patterning where the tantalum-based barrier is completely removed with a silica-abrasive slurry that is similar to an oxide slurry. A key objective is to relate the CMP removal rate and post-CMP surface condition to OSG film composition and CMP process variables. More specifically, we compare OSG and SiLK removal rates using copper and oxide slurries, with extensive results relating removal rate to oxide slurry pH. Surface measurements have been made following CMP to quantify physical and chemical changes to the OSG and SiLK films. In addition, the effects of film aging in the fab ambient and the variation of slurry acid additives are examined to further understand the chemical interactions involved between the CMP slurry and the OSG and SiLK films. The chapter concludes with a extensive discussion of copper damascene patterning issues with these OSG ILDs.

## 5.1 EFFECT OF FILM CARBON CONTENT

CMP of blanket films with 30 second polishing time was performed using an Applied Materials Mirra CMP system. The first set of experiments used five slurries to polish three OSG films with different properties resulting from different deposition conditions. The first four slurries are KOH-based silica slurries, the first as recommended by the manufacturer for oxide CMP and the others with propionic acid added to lower the pH from 10.8 to 9.5, 7.7 and 6.0 while maintaining the solids concentration at 3.5 wt%. The fifth slurry is an alumina slurry with QCTT1010 chemistry. The three types OSG films were deposited by PECVD with varying density (between 1.15 and 1.43 g/cm$^3$) and varying carbon content (between 20 and 28 at%). The five slurries are described in detail in Table A.4 and the OSG film properties are given in Table A.2. Removal rate, surface roughness, and chemical composition are discussed as a function of carbon content and film age.

### 5.1.1 Removal Rate in Oxide Slurries

Since OSG films are oxide-based, slurries for oxide with silica abrasives were used as a starting slurry. The oxide slurry is initially adjusted by modifying the pH from a starting value of ≈10.5, common for commercial oxide slurries. For comparison, the removal rate of SiLK, described extensively in Chapter 4, is also presented.

The removal rate of OSG in oxide polishing slurries over a wide range of pH is shown in Figure 5.1. The plot compares the removal rates of Applied Materials BlackDiamond BD1, Novellus Coral C1, and Novellus Coral C2 with PECVD tetra(ethylorthosilicate) glass (TEOS SiO$_2$) and SiLK polymer. The stock slurry used for blanket film experiments is a diluted SS11 potassium hydroxide (KOH) based slurry from Cabot Corporation, with the slurry pH adjusted with propionic acid.

Since the SS11 slurry is designed for the CMP of SiO$_2$, the TEOS oxide removal rate is expected to be higher than the OSG and SiLK removal rates at all pH levels. In particular, the carbon content in the OSG and SiLK suppresses the chemical reaction of the slurry with these films, resulting in a lower removal rate. The low-κ removal rate varies inversely with film carbon content, resulting in the following removal rate comparison: TEOS SiO$_2$ > BD1 > C2 > C1 >> SiLK. The SiO$_2$ polishing slurries have almost zero SiLK removal rate due to the SiLK bonding structure. The aromatic carbon rings in the polymer strongly resist the chemistry of the SiO$_2$ slurries.

Reducing the slurry pH decreases the OSG removal rate, due to reduced free OH$^-$ reactant in the slurry. Such a decrease in removal rate with pH has been observed when polishing silsesquioxane materials [5.1]. It appears that increasing the acidity of the slurry (and thus decreasing the free OH$^-$) decreases the ability of the slurry to hydrolyze and weaken the OSG film surface. In the case of C1 and C2, the decrease in removal rate with slurry pH follows the same trend as TEOS SiO$_2$ removal rate. In contrast, the BD1

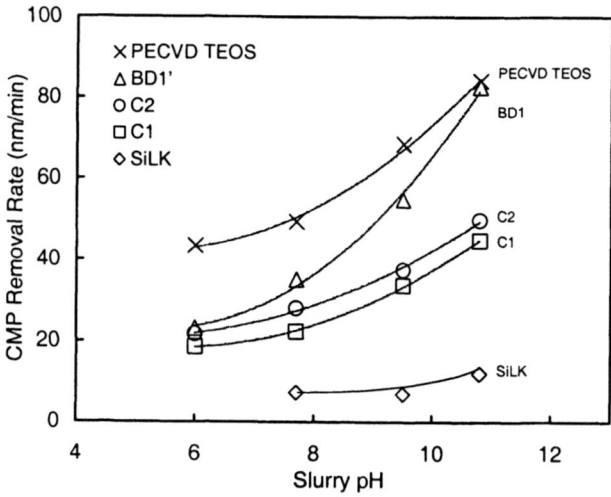

*Figure 5.* CMP removal rates for PECVD TEOS $SiO_2$, BD1, C1, C2, and SiLK with slurries A-D, silica slurries designed for silicon dioxide CMP. Slurry pH adjusted with propionic acid. P = 14 kPa, $V_{relative}$ = 1.4 m/s.

removal rate is as high as TEOS removal rate at a pH of 10.8, but as low as C2 removal rate at a pH of 6.0. The lower carbon content in the BD1 film had a significant impact on the variation of CMP removal rate with slurry pH. Since Si-O bonds are much more reactive in the oxide polishing slurries than Si-C bonds, an increase in the Si-O bond concentration of 5 to 10 percent greatly increases the ability of the oxide slurry to remove the BD1 film.

Two sets of removal rate data were measured for each OSG film, the first for a film age of approximately two weeks and the second for a film age of approximately two months. The films were aged uncovered in the fab ambient at low relative humidity and constant temperature. Therefore, it is possible that some water was absorbed in them. As the OSG films age, the CMP removal rate in slurries A-D increases as shown in Figure 5.2.

Figure 5.2(a) shows that the removal rate of BD1 increases with film age, nearly doubling when polished at a slurry pH of 10.8. The humidity in the fab ambient, however low, may cause water absorption into the OSG film and hydrolysis of structural bonds. Figures 5.2(b) and 5.2(c) show that the removal rates of C2 and C1 also increase with film age, but to a lesser degree than BD1. BD1 appears to be more susceptible to water absorption due to lower carbon content. C2 and C1 are more resistant to water absorption and hydrolysis reaction in the fab ambient due to higher carbon content.

Figure 5.3 shows that the refractive index of the films (measured by spectroscopic ellipsometry) is affected by aging these films. When the C1 and C2 films are initially measured, their RI measure 1.422 and 1.451

respectively; following CMP, their RIs decrease to 1.416 and 1.448, respectively. This slight reduction in RI may reflect the removal of a top layer of the OSG film which contains a small amount of absorbed water. When the films age in the fab ambient for 2 weeks following CMP, the RI for both films increase to their previous values. C1 contains more carbon that C2, resulting in less water absorption and a smaller reduction in RI due to CMP.

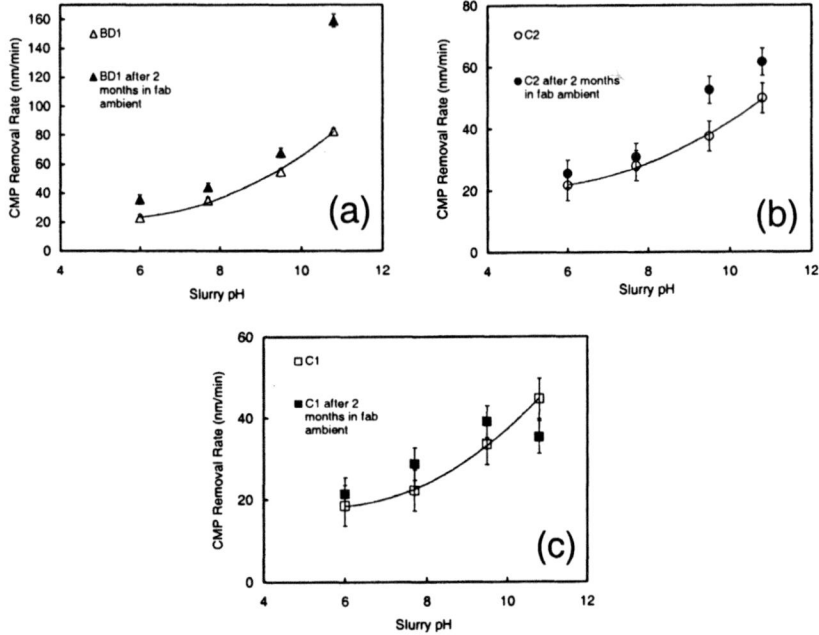

*Figure 5.2.* Plots of removal rate versus slurry pH for (a) BD1, (b) C2, and (c) C1, showing the effect of storage in the fab ambient. P = 14 kPa, V = 1.4 m/s

A third test compared the effect of the type of organic acid used to alter the slurry pH. Slurry A was adjusted with either propionic or citric acid to compare the effect of the acid type on the OSG removal rate. Both sets of silica slurries were adjusted to the same pH values immediately prior to CMP. Figure 5.4 shows the results of CMP tests on BD1, C2, and SiLK. Again, we observe the inverse trend of removal rate with film carbon content, as the removal rate follows BD1 > C2 >> SiLK. But for each film, the removal rate is higher when using the slurry containing citric acid. Citric acid (2-hydroxypropane-1,2,3-tricarboxylic acid, $C_6H_8O_7$) is a larger molecule than propionic acid (carboxyethane, $C_3H_6O_2$), and has three carboxylic acid groups per molecule compared to one carboxylic acid group per propionic acid molecule. Therefore, a smaller volume of citric acid is required to lower the slurry pH.

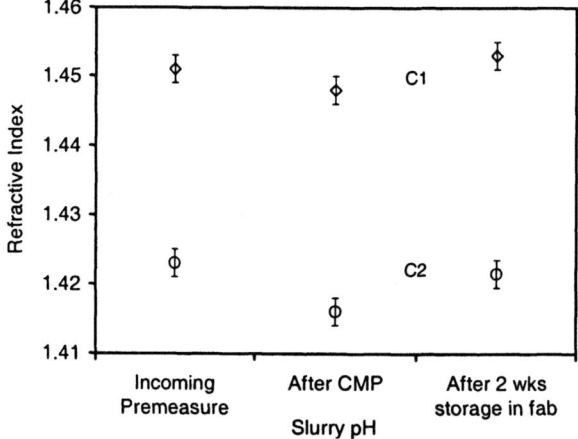

*Figure 5.3* Change in refractive index of C2 and C1 due to water absorption before and after CMP. Films polished using slurry A for 30 sec with P=14 kPa and V=1.4 m/s

Note that the different acids affect the CMP removal rate of SiLK (Figure 5.4(c)), even though SiLK does not react chemically with the basic slurry (and thus does not have as large a removal rate as the OSG films). The increase in the OSG and SiLK removal rates may be caused by more adequate slurry wetting due to the highly polar citric acid molecules adsorbing on the hydrophobic surface of both types of materials. Citric acid may behave as a surface-active agent that increases the CMP removal rate of the hydrophobic OSG and SiLK films by increasing slurry wetting, similar to results presented in the literature for CMP/surfactant interactions with other hydrophobic low-κ materials [5.2, 5.3].

## 5.1.2 Removal Rate in Copper Slurries

The CMP removal rate of BD1 and C2 with a slurry designed for copper CMP is shown in Figure 5.5. Two different pressure/velocity combinations were used to represent two possible recipes for copper removal. Very low OSG removal rate, on the order of 10 nm/min, was obtained for both sets of tool variables investigated. The acidic slurry chemistry does not rapidly remove OSG under the CMP conditions used. At higher CMP pressure, the removal rate of BD1 increases and the removal rate of C2, which has a higher carbon content, remains nearly constant. Lower carbon content may make the film more sensitive to CMP pressure, although both films have the same κ (~ 2.85). Since BD1 has 10% less carbon, it is reasonable to assume that the reduction in dielectric constant is due to a small-scale film porosity. This porosity may cause the BD1 film to be more susceptible to pressure.

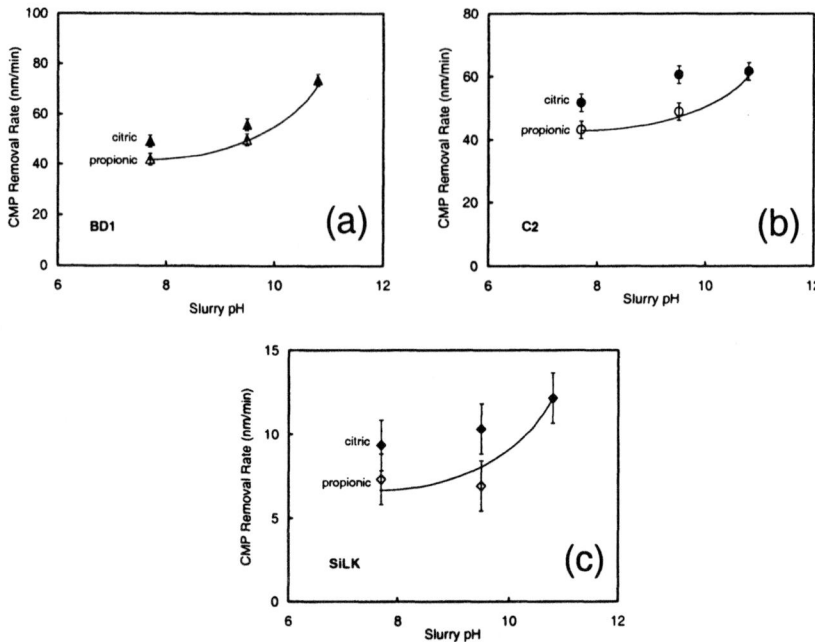

*Figure 5.4.* A comparison of two organic acids used to alter slurry pH, and their effect on the direct CMP removal rates for (a) BD1, (b) C2, and (c) SiLK films. P = 14 kPa, V = 1.4 m/s.

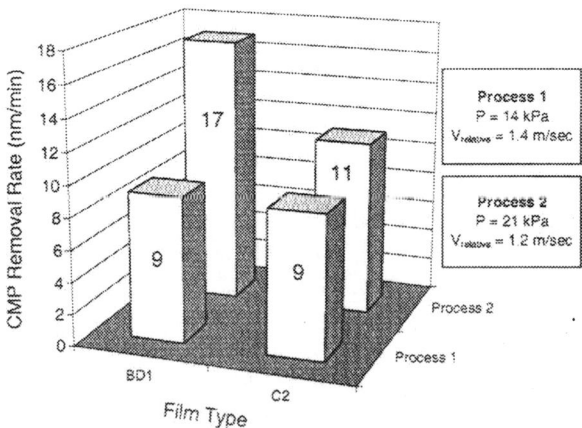

*Figure 5.5.* CMP removal rates for BD1 and C2 with slurry E, an alumina slurry designed for copper CMP. BD1 is more sensitive to increased pressure.

## 5.2 SURFACE ROUGHNESS

The surface roughness of OSG films was measured using atomic force microscopy (AFM) both before and after CMP. An interesting benefit of the CMP process was observed when polishing the BD1 film. After deposition and curing, the BD1 films used had high surface roughness due to hills of deposited material, as shown in Figure 5.6(a). These hills are a result of the deposition process, where the reactant plasma is adjusted to decrease the mobility of the product during deposition. The results are (i) clumping of deposited material on the wafer surface that can reach maximum heights of ~ 50 nm and (ii) high surface roughness (6.9 nm RMS).

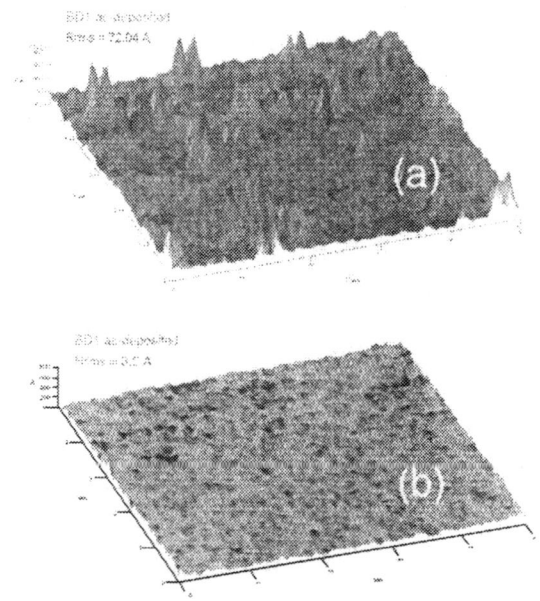

*Figure 5.6:* 5 μm x 5 μm AFM scans of (a) BD1 before CMP and (b) BD1 after 30 seconds CMP with slurry B

Following 30 seconds of CMP in slurries A-D, the surface hills are eliminated, and a smooth surface is obtained (shown in Figure 5.6(b)). The post-CMP RMS roughness for BD1 is 0.35 nm, with the highest feature measuring 1.9 nm. Very few surface scratches are observed following CMP of BD1 with slurry A.

Table 5.1 contains AFM data on surface roughness for BD1, C2 and SiLK (for comparison purposes) polished in different slurries. Due to the deposition conditions of C2, the film is an order of magnitude smoother than

BD1 after deposition (0.72 nm RMS, with a maximum feature height of 5 nm, compared to 6.94 nm RMS, with a maximum feature height of 47 nm). Following CMP with slurry A, the RMS roughness of C2 is reduced to 0.15 nm, with a maximum feature height of 0.5 nm, while comparable values for BD1 are 0.35 nm and 1.9 nm, respectively. Thus, the two OSG films polished in slurry A have improved roughness by a factor between 5 and 20, depending on the chosen metric and the initial film properties.

However, very shallow surface pits have been observed following the CMP of BD1 and C2 with slurry A, with the deepest pits measuring 1.4 nm. Similar defects have been observed following the polishing of single-crystal silicon wafers, and have been attributed to structural defects in the silicon lattice [5.4]. The pits may represent dimples left in the OSG surface due to abrasive impact, or may indicate selective removal of the film at areas of the surface which have lower carbon content after deposition.

*Table 5.1.* RMS roughness values measured by AFM before and after CMP in various slurries. Slurry pH adjusted by the addition of propionic acid

|  | Pre-CMP RMS Roughness (nm) | Post-CMP RMS Roughness (nm) | Comments |
|---|---|---|---|
| BD1 Slurry A | 6.94 | 0.35 | pre-CMP max feature = 47 nm; post-CMP max feature = 1.9 nm |
| C2 Slurry A | 0.72 | 0.15 | pre-CMP max feature = 5 nm; post-CMP max feature = 0.5 nm; post-CMP small pits observed max; pit depth = 1.4 nm |
| C2 Slurry D | 0.72 | 0.41 | post-CMP large pits observed; max pit depth = 50 nm |
| SiLK Slurry A | 0.45 | 1.6 | post-CMP max scratch depth 3.2 nm |
| SiLK-Cu CMP Slurry | 0.45 | 2.0 | post-CMP many small scratches observed |

Increased surface roughness is observed as the slurry pH is decreased below 7.0. Following CMP with slurry D (pH = 6.0), the RMS roughness of C1 measures 0.41 nm, and deeper pits are observed in the low-κ material surface. The maximum pit depth increases to 50 nm, and slurry residue is present near the pits. We attribute the less desirable surface to the instability of slurry D at a pH of 6.0. At this pH, the colloidal suspension of the silica particles is degraded compared to slurry A (pH 10.8), resulting in more agglomeration and settling of slurry particles. The decreased suspension of abrasive particles causes both damage to the OSG and slurry residue that is difficult to remove from the OSG surface.

Figure 5.7 shows a 5 μm x 5 μm AFM scan of the SiLK polymer surface following CMP in slurry A, for comparison to Figure 5.6(b). The post-CMP RMS roughness of SiLK is 1.6 nm following CMP, which compares very well with literature values for the roughness measured after polishing SiLK in copper slurry with high (~300 nm/min) removal rate [5.5]. The SiLK material is much softer than $SiO_2$ [5.6], resulting in the roughening of the SiLK surface by the CMP slurries. Although nanoscratches are much more prevalent in SiLK than in OSG, the SiLK surface roughness measured is considered acceptable for Damascene processing [5.7]. The surface roughness is actually quite low, considering the low removal rate of the CMP process of SiLK in oxide slurry. Low removal rate processes often are correlated with appreciably roughened and scratched soft surfaces, but this is not observed for the CMP conditions used (14 kPa, 1.4 m/s).

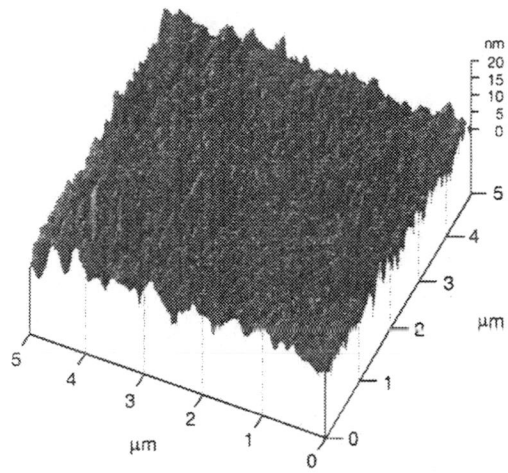

*Figure 5.7:* 5 μm x 5 μm AFM scan of SiLK after 30 seconds CMP with slurry A

Several of the post-CMP defects observed on the OSG films are shown in Figure 5.8, along with SiLK defects for comparison purposes. Figure 5.8(a) shows some shallow surface pits in the C2 film surface following CMP with slurry A, while Figure 5.8(b) shows slurry residue surrounding a deeper pit in C2 following CMP with slurry D. Pit and residue defects on the OSG surfaces increase in number as the slurry pH is decreased below 7.0. Figures 5.8(c) and 5.8(d) compare the smooth OSG surfaces following a high removal rate CMP process with rougher SiLK surfaces following a low removal rate CMP process. Figure 5.8(c) illustrates the shallow surface scratching that is observed for soft materials due to CMP. Figure 5.8(d) is an enlargement of the residue near the center of Figure 5.8(c), showing that some slurry residue remains at the bottom of the shallow defect. This type of blistering or ripping defect has been observed when polishing polymer materials, and is attributed to physical abrasion from the slurry particles and pad [5.8]. Figure 5.8 contrasts the surface quality of the OSG CMP process which has high removal rate due to surface-altering slurry chemistry with the quality of the SiLK CMP process, which has low removal rate due to negligible surface-altering chemical reaction.

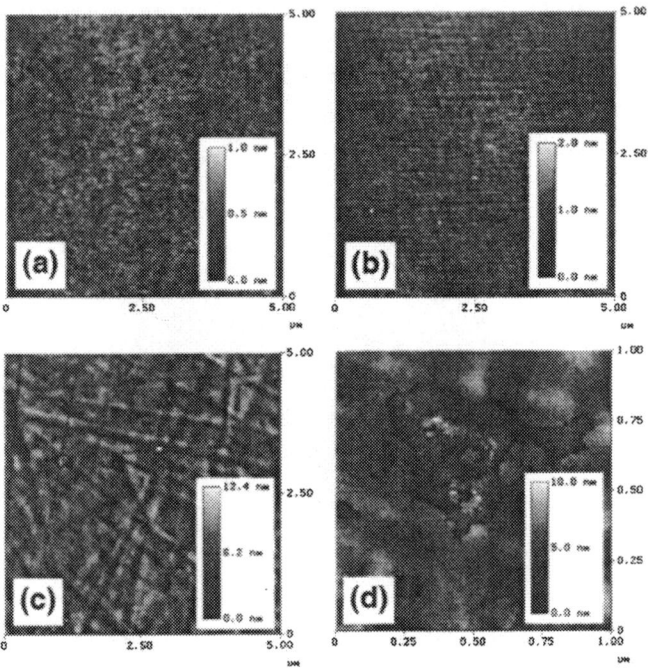

*Figure 5.8:* Top view AFM scans showing (a) C2 shallow surface pits, (b) C2 abrasive near pits after CMP at pH = 6.0, (c) SiLK surface scratching, and (d) magnified SiLK scan showing slurry residue remaining within defect.

## 5.3 SURFACE AND BULK FILM CHEMISTRY

The chemical composition of the OSG films was measured using X-ray photoelectron spectroscopy (XPS) and Fourier-transform infrared spectroscopy (FTIR) to determine the effect of CMP with slurries A-D.

These measurements are useful in establishing the chemical reactions that occur during CMP and their impact on film structure, chemistry, and electrical properties.

### 5.3.1 XPS Surface Results

XPS results for BD1 and C2 are shown in Figures 5.9(a) and (b) and Table 5.2. Figure 5.9(a) shows the carbon binding energy window for BD1 before and after CMP in the colloidal KOH-based silica slurries with varying pH (A:10.8; B:9.5; C:7.7, and D:6.0). To account for sample charging during measurement, each spectrum has been normalized to the main silicon peak energy of 101.6 eV, which corresponds to the bond energy of Si-O. The binding energy shifts for each peak are shown on the plot labels in parentheses. Any large changes in the peak location and peak area before and after CMP reflect changes in the bonding and composition of the film.

Table 5.2. XPS results for BD1 and C2 before and after CMP. CMP does not measurably change film silicon, oxygen, or carbon content

|  | Slurry | C (1s) at% | N (1s) at% | O (1s) at% | Si (2p) at% |
|---|---|---|---|---|---|
| BD1 | Unpolished | 22.5 | 3.0 | 42.7 | 31.8 |
|  | A (pH=10.8) | 22.4 | 2.4 | 44.6 | 30.7 |
|  | B (pH=9.5) | 22.7 | 2.4 | 45.0 | 29.9 |
|  | C (pH=7.7) | 23.3 | 2.5 | 43.3 | 31.0 |
|  | D (pH=6.0) | 23.8 | 2.4 | 43.4 | 30.4 |
| C2 | Unpolished | 31.3 | 2.2 | 38.7 | 27.8 |
|  | A (pH=10.8) | 29.2 | 2.1 | 41.5 | 27.2 |
|  | B (pH=9.5) | 31.0 | 1.7 | 39.9 | 27.3 |
|  | C (pH=7.7) | 30.3 | 1.9 | 39.9 | 27.9 |
|  | D (pH=6.0) | 30.5 | 1.7 | 40.9 | 26.9 |

No large changes in peak location or shouldering (peak splitting into multiple carbon bonded constituents) were observed in the BD1 film. The Si-C carbon peak location in Figure 5.9(a) remains within an energy range of 286.5 eV – 287.5 eV following CMP, suggesting that the carbon present in the surface layer of the film is bound to silicon with the same chemical structure as the unpolished film. No noticeable peak shoulders appear at higher or lower binding energies, suggesting that C-O and C=O bonds are not formed due to CMP.

Table 5.2 shows the atomic content of C, Si, O, and N in BD1 before and after CMP. The elemental composition as measured by the peak areas in

Figure 5.9(a) of BD1 remains constant before and after CMP, within ± 2 atomic percent. This suggests that the potassium hydroxide (KOH) based slurry does not selectively attack the Si-C bonds in the film, which would result in a lower carbon content in the surface layer of the film following CMP. Conversely, the slurry does not selectively etch away Si and O, leaving a carbon-rich layer at the film surface.

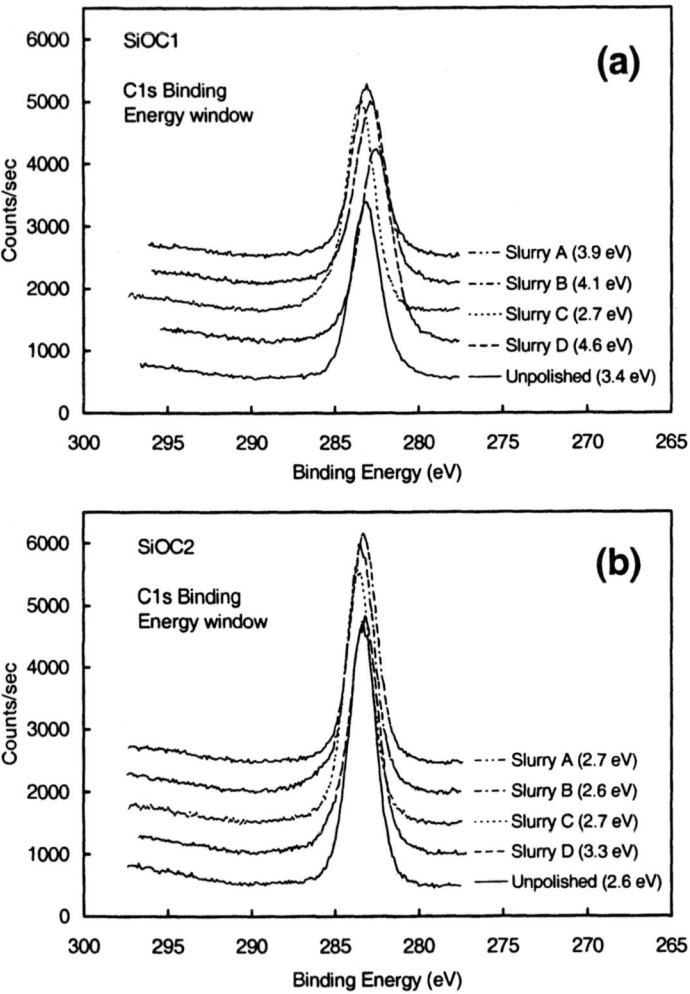

*Figure 5.9:* XPS C1s binding energy window for (a) BD1 and (b) C1 films before and after CMP in slurries A-D (pH adjusted by propionic acid). No change in surface composition is observed after CMP

## CMP OF ORGANOSILICATE GLASSES

Similar results are shown for C2 in Figure 5.9(b). Here the peak heights are larger than in BD1 due to higher carbon content. After CMP, the Si-C carbon peak location remains within an energy range of 286.0 eV – 286.5 eV, suggesting that the chemistry of the carbon bonds in the C2 surface is unchanged by the CMP process. The absence of carbon shouldering peaks is evidence that the carbon in the C2 structure does not become oxidized by the CMP slurries. Table 5.2 shows that the atomic concentrations of C, Si, O, and N in C2 remain constant before and after CMP, again within ± 2 atomic percent. While the chemical composition of the C1 film was not measured, its structure is very similar to C2, so that similar characteristics are anticipated.

### 5.3.2 FTIR Bulk Profiling Results

The bulk chemistry of the OSG films was measured by transmission Fourier-transform infrared (FTIR) analysis in the spectrum of 400 to 4000 $cm^{-1}$. Adsorption at certain energies is related to chemical bonds and atomic concentrations that are similar to the XPS measurements discussed previously. However, the transmission FTIR beam passes completely through both the silicon wafer and the OSG film, providing information about the chemical bonding and structure in the film bulk rather than at the film surface.

Figure 5.10 shows FTIR traces for BD1, C2, and C1 before and after CMP in slurry C, with similar results obtained with slurries A, B, and D. For all slurries, the pre-CMP and post-CMP measurements are nearly identical, suggesting that the CMP process does not chemically or structurally change

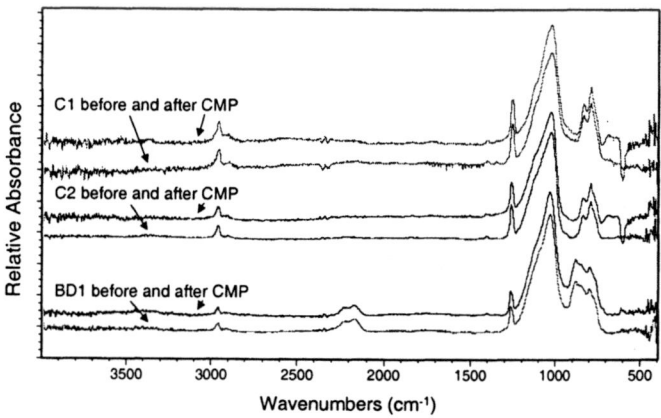

*Figure 5.10.* Transmission FTIR spectrum of OSG films before and after 30 sec CMP with slurry C. Spectrum obtained by subtraction of silicon substrate reference

the bulk of the films (to the resolution limits of FTIR with 0.5 μm thick films). Thus the slurry does not penetrate, react, or absorb through the entire 500 nm depth of the OSG. Figure 5.10 also shows that the CMP process does not introduce a substantial amount of water into the OSG films, which would be observed from a wide band of peaks in the range of 3400 to 3600 cm$^{-1}$. The structural bonds that are vital for good OSG dielectric properties (C-H$_3$ 2950 cm$^{-1}$, Si-H 2200 cm$^{-1}$, Si-CH$_3$ 1250 cm$^{-1}$, Si-O$_2$ 1050 cm$^{-1}$) do not change in concentration or character as the result of CMP.

## 5.4 COPPER DAMASCENE PATTERNING WITH OSG DIELECTRICS

The blanket film results described in the previous sections (and Chapter 4) develop trends and methods for integration of OSG (and other low-κ dielectrics) with copper metallization. Successful multilevel integration of low-κ and copper by damascene patterning depends on successful patterning, etching, and CMP planarization at the feature (~nm), die (~cm) and wafer (~20cm) scales. The capability of CMP for high-yield manufacturing is determined in large part by film carbon content (section 5.1), removal rate selectivity, adhesion, planarization length, and damage to the low-κ film in the form of scratch defects or surface roughness (section 5.2). Table 5.3 lists several criteria for successful low-κ integration.

The challenges in meeting the above criteria exist for each of the low-κ options discussed in Chapter 2. The following sections discuss low-κ issues in general, focusing specifically on the challenges of interconnect formation by single damascene patterning of carbon-doped glasses (OSGs).

*Table 5.3.* List of properties desired for successful integration of low-κ films (adapted from [5.9])

| Metric | Goal |
|---|---|
| Adhesion (to metal, self-adhesion) | Pass tape test after thermal cycles to 450 C |
| Gap-fill | No voids at 0.35 μm, aspect ratio = 2 |
| Planarization | > 80% (regional) |
| Etch rate | > 3 nm/s |
| Step coverage | > 80% |
| Reliability of metal, when surrounded by dielectric material | |
| Resistance to solvents and photoresist strippers | |
| Etch selectivity (oxygen plasma resistance) | |
| CMP compatibility | |

## 5.4.1 Hardmasks or Dielectric Cap Layers

The soft nature of low-κ dielectrics, as well as their high carbon content, often requires the use of a hardmask or dielectric cap during integration (described briefly in Sections 1.1 and 3.2.3). This hardmask serves as a sacrificial film above the low-κ IMD/ILD that allows patterning, etching, and the ability to stop the CMP process without directly abrading the soft low-κ film. The most popular hardmasks are silicon nitride or silicon dioxide. Both films have well-known properties and behaviors during etch and CMP. $SiN_x$ and $SiO_2$ are also resistant to the ashing and wet cleaning chemistries that are used to clean away photoresist following etch, that may attack the organic constituents of SiOC or polymers. SiOC and other polymers have been developed to meet the k = 2.7 interconnect technology node. However, when a hardmask is used in damascene patterning, the effective dielectric constant of the integrated film, which includes the dielectric contributions of the IMD/ILD and the remaining hardmask material, must be considered. Figure 5.11 shows the significant impact that a thin hardmask can have on the dielectric advantage of a low-κ film.

The figure shows that an integrated interconnect structure using SiOC at both ILD and IMD levels (k = 2.7/2.7) will have a dielectric constant of 3.91 (with $SiN_x$ cap) compared to 3.13 (with $SiO_2$ cap) and 2.7 with no cap. While no-cap is desired electrically, the impact of each integration step on the dielectric constant of an uncapped film needs to be considered. Damascene patterning involves photo-resist application, patterning, etching, metallization, and CMP, with multiple clean and anneal steps between processes. Figure 5.12 shows dielectric constant measurements performed after each integration step in damascene patterning of uncapped SiOC.

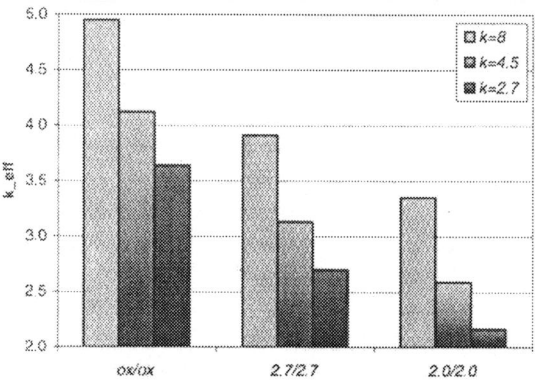

*Figure 5.11.* Effective dielectric constant calculated for various combinations of hard masks (κ = 2.7, 4.5 and 8) and interlevel/intralevel low-κ materials. Hardmask thickness is ~100 nm, while dielectric thickness is ~400 nm. Adapted from [5.10]

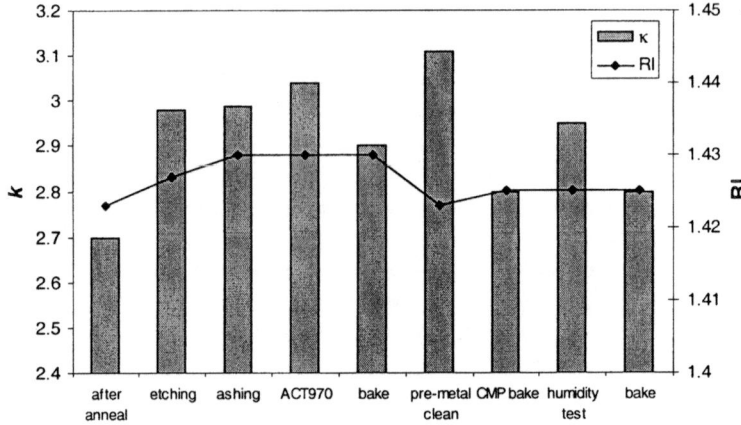

*Figure 5.12.* Changes of dielectric constant with single damascene integration steps. ACT970 is a wet chemistry used for photoresist clean-up. Adapted from [5.11]

Figure 5.12 illustrates the care that must be taken to optimize the dielectric constant advantage allowed by a low-κ ILD. Plasma and wet chemistry treatments have the most detrimental effect on the dielectric constant of SiOC, while bake steps can often be used to drive off silanol or absorbed water, restoring dielectric constant to within 4% of its initial value.

### 5.4.2 Low-κ Etching

Damascene patterning uses plasma etch chemistries to etch trenches or vias into the low-κ dielectric material. The ionized gases break the bonds of the low-κ films, creating volatile gaseous species and etching away the trench pattern. The etched channels are subsequently filled with liners and low resistivity copper. Fluorocarbon etch gases such as $CF_4$, $CHF_3$, and $C_2F_6$ are commonly used to etch silicon-dioxide dielectric. First-run etch experiments using $SiO_2$ etch chemistries with varying addition of oxygen plasma result in the low-κ trench etch profiles shown in Figure 5.13.

The similarity in molecular structure of SiOC allows the same basic etch plasma reactions as occur in $SiO_2$ removal. However, high carbon or hydrogen contents in the SiOC film limit the etch rate as the fluorocarbon etchants react with the organic content to form non-volatile polymers [5.13]. Hydrogen scavenges fluorine, reducing the etch rates of SiOC or other low-κ materials containing organics to levels below the etch rate of $SiO_2$. The etch rate can be increased by introducing a low content of oxygen plasma. However, an $O_2$ plasma will preferentially attack the organic content in the film, causing potentially detrimental effects to etch profile and dielectric constant of the dielectric exposed [5.12].

CMP OF ORGANOSILICATE GLASSES 113

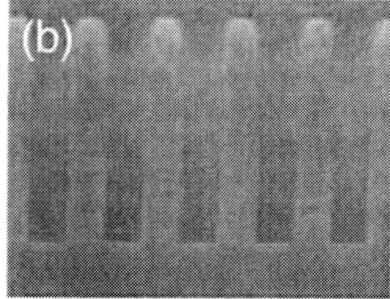

*Figure 5.13* – Sample etch profiles for (a) SiLK and (b) SiOC films using conventional fluorine/oxygen based plasmas. From [5.12].

### 5.4.3 CMP Integration

Once the SiOC low-κ film is patterned, etched, cleaned, and baked, the barrier and Cu conductor are deposited into the trenches and onto the dielectric field (the surface region between the trenches which have not been etched). The result of continuous-film barrier deposition and electro-deposition is a significant copper overburden (a layer of copper and liner deposited on the dielectric field) that is removed with CMP, as described in Chapter 3. One of the most difficult aspects of low-κ integration is the issue of adhesion between the films in the ILD/IMD stack. Figure 5.14 shows an example IMD (trench level) stack and the interfaces at which adhesion may become an issue.

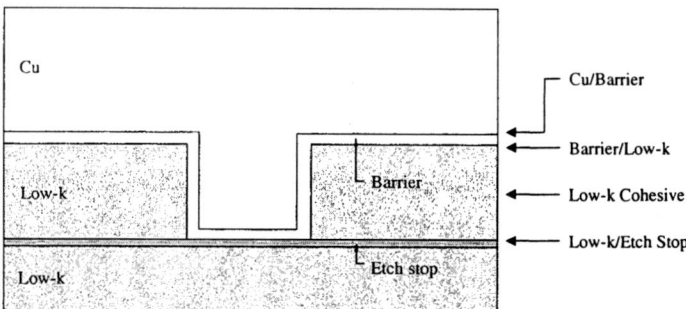

*Figure 5.14.* Schematic IMD stack, showing interfaces where the stack may be susceptible to adhesion failures during CMP.

The normal and shear stress imparted to the wafer surface during CMP can cause adhesive failure at an interface between layers (i.e. barrier/low-κ) or cohesive failure within a layer or material. The soft nature of most low-κ dielectrics causes them to be susceptible to cohesive failures during polish. Typically, the order of interface adhesive strength in a patterned structure follows the trend Cu/barrier > low-κ/etch stop > low-κ cohesive > barrier/low-κ. The weakest interface is typically between the barrier and the low-κ material [5.14]. This interface may be the weakest due to residue remaining on the low-κ surface following post-etch clean-up, or due to the hydrophobic nature of the low-κ material which reduces surface wetting during barrier deposition.

Adhesion is typically measured using the qualitative "tape peel test" or the more quantitative "modified edge lift-off test (m-ELT)", [5.15] which measures the cohesive fracture toughness of a material or the adhesive strength of an interface. These tests are generally performed on blanket films which have the minimum adhesive strength between layers, since patterned films have more surface area within the trenches to increase contact adhesion. Blanket tape tests reflect the adhesion between layers in large open field areas [5.16] between arrays or the adhesion of barrier to ILD during single-damascene via patterning, which typically has <<10% patterned area.

Adhesion can be improved by plasma treatment of polymer or SiOC dielectrics, due to a preferential etch of organic content and a slight roughening of the film surface. As mentioned, the hydrophobic nature of most low-κ dielectrics causes the barrier/low-κ interface to have the least adhesive strength. Hardmasks or cap layers can improve adhesion, at the cost of effective dielectric constant. An $SiO_2$-like cap has excellent adhesion to most barrier materials and is typically deposited by plasma-CVD, which will enhance the low-κ surface adhesion. Other options include spin-on or in-situ deposited adhesion promoters.

Once adhesion issues are overcome, the next challenge involves the variable polish rates or "selectivity" of the different materials in a low-κ interconnect level. The CMP of patterned films involves the simultaneous polishing of three (Cu, barrier, and dielectric) or possibly four materials, if a hardmask is used to cap the low-κ dielectric. Each material has a different chemical structure, ranging from electrochemically active (Cu) to hard and chemically inactive (Ta/TaN or Ti/TiN barriers), to soft and chemically inactive (polymers or high-carbon-content glass).

Barrier polish slurries are developed with removal rate properties that fall under two categories -- high selectivity slurries (HSS) and low selectivity slurries (LSS). An HSS has relatively high Cu and barrier removal rate, and low dielectric removal rate. An LSS is engineered to have nearly equal rates for Cu, barrier, and low-κ. The chemical components in a slurry can be tuned to increase the removal rate of one or more materials, or by using passivants to reduce the chemical attack of one or more materials. Figure 5.15 shows a plot of removal rate versus slurry additive for an SiOC and $SiO_2$ cap layer.

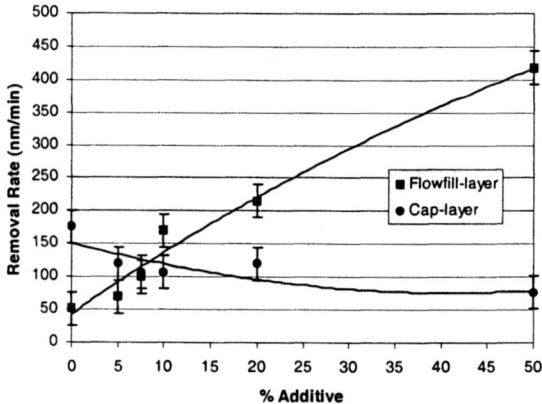

*Figure 5.15.* Removal rates of cap and the low-κ Flowfill (SiOC) layer as a function of the slurry additive. The removal rate selectivity is adjustable from about 3:1 to 1:5. From [5.17]

The removal rate selectivity observed in Figure 5.15 is adjusted as the slurry additive passivates the $SiO_2$ surface and makes the SiOC surface more hydrophilic, allowing more intimate slurry contact and abrasive removal. This type of selectivity adjustment is vital for successful integration of low-κ dielectrics, whose relatively soft and non-reactive structure can lead to significantly lower removal rates, with much higher abrasion or scratching of material. Low-κ materials such as SiLK are susceptible to scratching at low removal rates. Figure 5.16 shows that the roughness of the SiLK surface decreases rapidly as the removal rate increases with increasing KH phthalate concentration. Thus an HSS approach may result in undesired roughness and defectivity when used with hydrophobic polymers or SiOC.

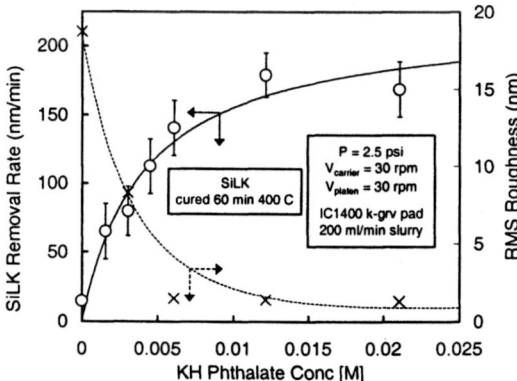

*Figure 5.16.* Inverse trends of post-CMP roughness with removal rate for blanket SiLK low-κ dielectric films.

HSS and LSS approaches are shown schematically in Figure 5.17. An HSS approach with low-dielectric constant materials may require either a hardmask/CMP cap or a post-barrier buff slurry to reduce scratching and roughness of the low-κ film. The challenge of an LSS approach comes in endpointing the barrier removal process -- selecting a timed polish that will clear remaining copper and barrier without overly thinning the IMD or ILD.

Damascene integration of SiOC or other low-κ dielectrics requires etch, clean, and CMP approaches that balance cap layers and slurry selections with the final desired effective dielectric constant and post-CMP film thickness. Main integration challenges include damage during etch in oxygen plasmas, water absorption during wet cleans, sufficient adhesion of layers to resist the forces of CMP, and surface defectivity which is often correlated with low CMP removal rate. Selection of a high-selective or low-selective CMP slurry can achieve the desired post-CMP film properties, while ensuring complete removal of residual copper and barrier. Cu damascene with SiOC dielectric can result in a dielectric constant advantage of 10% (with $SiO_2$ cap) to 23% lower effective dielectric constant, leading to faster, lower power interconnects.

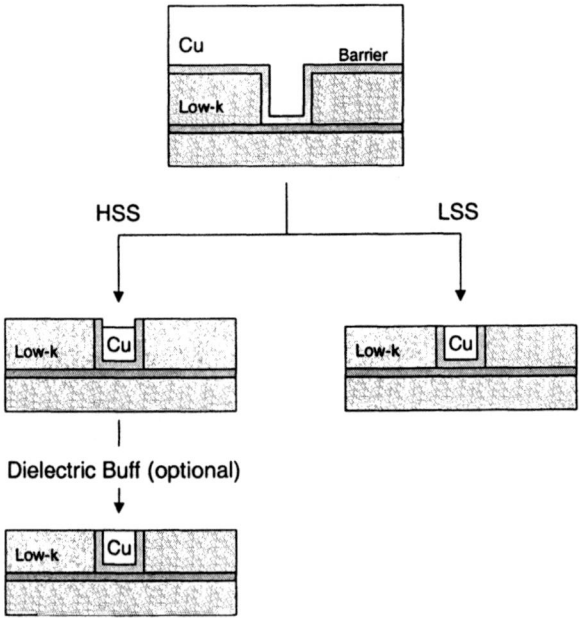

*Figure 5.17* – Schematic diagram of HSS (with optional dielectric buff) and LSS slurry single-damascene approaches.

## 5.5 SUMMARY

The removal rate for all OSG films is lower than or equal to the removal rate of TEOS oxide since the carbon content of the OSG suppresses the slurry chemical reaction with the Si-O bonds in the film. OSG is removed more rapidly in oxide polishing slurries than in copper polishing slurries due to the slurry pH and corresponding slurry reactivity. OSG is removed more readily in a basic medium, as removal rate increases with the concentration of $OH^-$ ions in the slurry. The OSG with highest carbon content has the lowest removal rate.

The post-CMP OSG surfaces are smooth over a range of basic pH values, but at neutral pH the decline in the abrasive particle suspension results in higher roughness and increased slurry residue following CMP. Such agglomeration of abrasive particles in the slurry also results in a larger number of post-CMP surface defects at slightly acidic pH. Surface chemical analysis before and after CMP shows that the reaction of the slurry with the OSG surface does not change the surface chemistry or composition. The slurry weakens structural bonds on the film surface and, in combination with shear and abrasion, removes Si, O, and C at identical rates. Slurry penetration and reaction is limited to the film surface, as the chemical composition of the film bulk is not altered.

OSG CMP in oxide slurries is quite feasible, with reproducible removal rate and low film damage. This enables the possibility of ultrathin copper barriers or removal of oxide caps from the Cu/low-κ structures, which is particularly important with scaling constraints as feature sizes approach 50 nm. The experimental data presented in this section is used n Chapter 6 to develop a mechanism for low-κ materials which includes polymers and doped glasses.

## 5.6 REFERENCES

[5.1] L. Forester, D. K. Choi, R. Hosseini, United States Patent No. 5,952,243, Sept. 1999.
[5.2] J. M. Neirynck, S. P. Murarka, R. J. Gutmann, in: T.-M. Lu, S. P. Murarka, T.-S. Kuan, C. H. Ting, *Low-Dielectric Constant Materials - Synthesis and Applications in Microelectronics*, San Francisco, USA, April 17-19, 1995, Materials Research Society Symposium Proceedings, **381**, 229 (1995).
[5.3] D. Permana, S. P. Murarka, M. G. Lee, S. I. Beilin in: R. Havemann, J. Schmitz, H. Komiyama, K. Tsubouchi, *Advanced Metallization and Interconnect Systems for ULSI Applications in 1996*, Boston, USA, October 1-3, 1996, Proceedings of Advanced Metallization and Interconnect Systems for ULSI Applications in 1996, 539 (1997).
[5.4] D. Gräf, M. Suhren, U. Lambert, R. Schmolke, A. Ehlert, W. von Ammon, P. Wagner, *J. Electrochem. Soc.*, **145(1)**, 275 (1998).
[5.5] C. L. Borst, D. G. Thakurta, R. J. Gutmann, W. N. Gill, *J. Electrochem. Soc.*, **146(11)**, 4309 (1999).

[5.6]   C. L. Borst, W. N. Gill, and R. J. Gutmann, Int. Jrnl. of Chemical-Mechanical Planarization for On-Chip Interconnection, **1(1)**, 26 (2000).
[5.7]   R. DeJule, Semicond. Intl. **20(13)**, 54 (1997).
[5.8]   D. Towery and M. Fury, J. Elect. Mat., **27(10)**, 1088 (1998).
[5.9]   G. Maier, Prog. Polym. Sci. **26**, 3 (2001).
[5.10]  R.A. Donaton, B. Coenegrachts, M. Maenhoudt, I. Pollentier, H. Struyf, S. Vanhaelemeersch, I. Vos, M. Meuris, W. Fyen, G. Beyer, Z. Tokei, M. Stucchi, I. Vervoort, D. De Roest, K. Maex, *Microelectronic Eng.*, **55**, 227 (2001).
[5.11]  J.-H. Lee, N. Chopra, J. Ma, Y.-C. Lu, T.-F. Huang, R. Willecke, W.-F. Yau, D. Cheung, E. Yieh, *Mat. Res. Soc. Symp. Proc.*, **612**, D3.4.1 (2000).
[5.12]  I. Morey, A. Asthana, *Solid State Technology*, June (1999).
[5.13]  T. E. F. M. Standaert, P. J. Matsuo, S. D. Allen, G. S. Oehrlein, T. J. Dalton, J. Vac. Sci. Technol. A **17(3)**, 741 (1999).
[5.14]  F. Lanckmans, S. H. Brongersma, I. Varga, H. Bender, E. Beyne, K. Maex, *Mat. Res. Soc. Symp. Proc.*, **612**, D1.4.1 (2000).
[5.15]  E. O. Shaffer II, F. J. McGarry, L. Hoang, *Polym. Sci. Eng.*, **36**, 2381 (1996).
[5.16]  E. S. Lopata, L. Young, J. T. Felts, *Mat. Res. Soc. Symp. Proc.*, **612**, D5.3.1 (2000).
[5.17]  E. Hartmannsgruber, G. Zwicker, K. Beekmann, *Microelectronic Eng.*, **50**, 53 (2000).

## Chapter 6

# LOW-κ CMP MODEL BASED ON SURFACE KINETICS

The discussion of removal rate, surface topography, chemical composition, and structural analysis in the previous chapters has been used to develop a physically-based conceptual model for low-κ CMP. The model assumes an altered-layer surface mechanism approach to represent the CMP of BCB, SiLK, and OSG materials and provide a generic understanding of the CMP process for other materials. The model also assumes a desirable low-κ CMP process where atomic-scale smoothness is achieved through an appropriate chemical-mechanical balance. The model does not account for deviations from an ideal CMP process, i.e. defect types such as scratching or peeling of the softer low-κ materials. Experimental results using SiLK polymer have been used to test and validate the basic approach and the model mechanism as an adequate representation of the complex CMP process.

Chemical engineers often use Langmuir-Hinshelwood (L-H) kinetics to describe the fundamental reactions that occur between species at a surface. The subsections of this chapter describe (1) experimental efforts to isolate the surface reaction that occurs during CMP, (2) incorporation of the surface reaction into an L-H kinetic mechanism, (3) testing the surface mechanism to determine the most important rate-controlling steps, and (4) simplification of the CMP surface mechanism to analyze the most important fundamental processes that occur during low-κ CMP. The L-H surface mechanism provides a boundary condition for solving the coupled fluid-mechanics and mass-transport based CMP model developed by Sundararajan[6.1] and Thakurta[6.2],[6.3] and enables one to determine the important steps in the process. When the most important process steps are emphasized, a simplified two-step model follows from the five-step mechanism, revealing additional aspects of interest in low-κ CMP. The two models described (complete L-H mechanism and simplified two-step mechanism) can be used to select CMP process parameters such as slurry composition, pressure, and velocity that are compatible with the chemical and mechanical properties of the ILD material chosen for multilevel integration. Such a balanced CMP process is more likely for polymers with relatively strong mechanical properties and moderate chemical reactivity. The SiLK, BCB, and OSG discussed in Chapter 5 meet these requirements, but other low-κ dielectrics may not.

## 6.1 ISOLATION OF THE CHEMICAL EFFECTS IN SILK CMP

Understanding the balance of chemical reaction and mechanical abrasion is central to the development of an effective CMP model. To this end, an effort has been made to isolate the slurry component that results in the rapid SiLK removal observed in Chapter 4.2. Observed rates were low in nitric acid slurries and high in QCTT1010 slurries, suggesting a distinct difference in surface mechanism. Follow-up experiments tested SiLK static etch and CMP removal rates in several different slurries.

The slurry mixtures and results are shown in Table 6.1. All slurries result in zero static etch rate of SiLK. This suggests that even when a surface-altering reaction occurs, the chemical attack does not effectively etch or "dissolve" the SiLK material. Rather, the slurry chemistry selectively breaks bonds in the SiLK surface, creating a weakened layer. The shear and abrasion of the CMP process are required to remove this altered reacted layer.

Table 6.1. Slurries, etch rates, and CMP rates used to isolate the effects that are important for SiLK removal.

|  | SiLK Wet Etch Rate (nm/min) | SiLK CMP Removal Rate (nm/min) | Copper CMP Removal Rate (nm/min) |
|---|---|---|---|
| DI water + 0.05 μm $Al_2O_3$ | 0 | 2 |  |
| DI water + $HNO_3$ + 0.05 μm $Al_2O_3$ | 0 | 10-15 | 150 – 200 (w/BTA) |
| DI water + $H_2O_2$ + 0.05 μm $Al_2O_3$ | 0 | 10-15 |  |
| DI water + QCTT1010 + $H_2O_2$ + 0.05 μm $Al_2O_3$ | 0 | 150 | 150 – 200 |
| DI water + QCTT1010 + 0.05 μm $Al_2O_3$ | 0 | 215 | ~ 50 |

The first slurry, a mixture of DI water and alumina abrasive, was examined to determine the amount of purely mechanical removal of SiLK under the CMP conditions. In this slurry, mechanical removal occurs on a very low scale, comparable to the removal rate of BCB and SiLK in slurry 1 (Table A.3, and Figures 4.1, 4.2). Similar results were measured for a slurry consisting of DI water, $H_2O_2$, and alumina. Again the SiLK polymer was not removed at a substantial rate, suggesting that the oxidizer alone does not accelerate the removal of SiLK.

The addition of QCTT1010 chemical to the slurry results in a significant increase in the SiLK removal rate. This rate is higher when there is no peroxide added to the slurry, suggesting that the QCTT1010 and $H_2O_2$ may have competing chemical reactions at the SiLK surface, and the oxidation of the polymer surface results in passivation that deters the structure-weakening reaction of the QCTT1010 with the SiLK bonds. This is an interesting result, since previous work has suggested that an oxidation/reduction reaction enables the CMP of some polymers[6.4]. Copper metal is removed in the QCTT1010 and $HNO_3$ slurries at a rate of 150 – 200 nm/minute. Thus the selectivity of SiLK to Cu CMP removal rate varies with the slurry chemistry.

Surface chemical analysis was used to measure the surface chemistry of the SiLK samples following static etch and CMP tests, to observe chemical changes in the SiLK surface due to chemistry or abrasion. Figure 6.1 shows that static dip and CMP tests using the slurry consisting of DI water, $H_2O_2$, and oxidize the SiLK surface. –C-O- shouldering peaks are clearly illustrated in the figure, suggesting that the peroxide component oxidizes SiLK bonds.

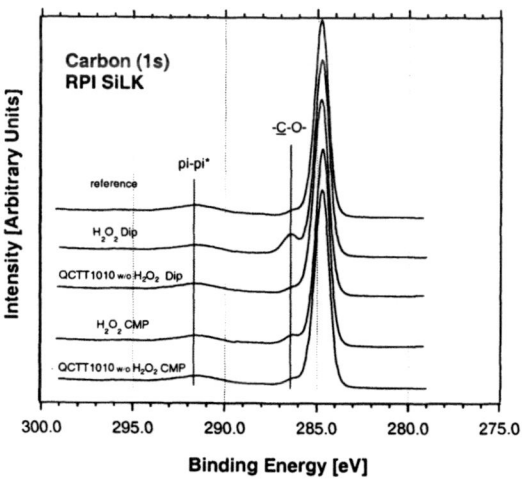

*Figure 6.1.* XPS spectrum of SiLK polymer following static etch and CMP removal tests in hydrogen peroxide and QCTT1010 chemistry.

Although the QCTT1010 chemical is responsible for the reaction that enhances SiLK removal, it does not oxidize the SiLK surface. But the QCTT1010 does obviously alter the SiLK material by some chemical means, allowing the shear and abrasive of the slurry to remove the SiLK material at a high rate. An additional experiment has been performed to examine the composition of QCTT1010 required to achieve SiLK material removal. The last slurry in Table 6.1 has been diluted while keeping abrasive content constant, to determine how diluting the active slurry chemistry affects the SiLK CMP removal rate. The result is shown in Figure 6.2.

Figure 6.2 shows that the QCTT1010 concentration has a profound effect on the SiLK CMP removal rate. The removal rate is linearly dependent on concentration at volume percentage less than 30%, and saturates above 30% at a maximum value of approximately 250 nm/min. The linear increase of removal rate, approaching an asymptote indicates the presence of a surface-chemistry dominated reaction mechanism, several of which have been described by Smith[6.5]. The following sections describe further slurry chemistry experiments and the modeling approach that is used to represent the surface reaction between SiLK and the QCTT1010 slurry chemistry.

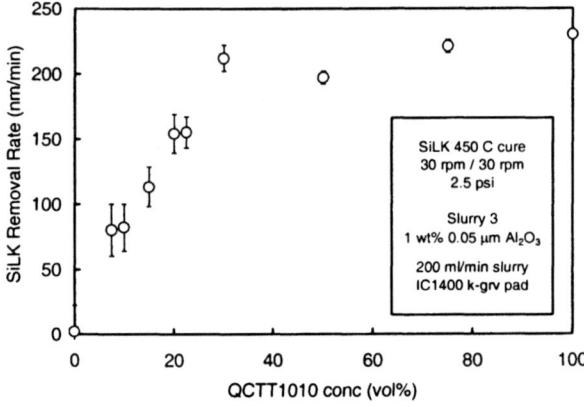

*Figure 6.2.* SiLK removal rate as a function of QCTT1010 slurry dilution. Abrasive content remains constant.

## 6.2 CMP WITH MODEL SILK SLURRIES

Two slurry chemistries have been examined to isolate the specific chemical agent that results in high SiLK removal rate. Table 6.2 lists the slurry components. Both slurries consist of DI $H_2O$ and phthalic acid -- a small organic acid with two neighboring carboxyl groups -- which is an excellent chelating agent for copper. A chelating agent is a substance whose molecules can form several coordinate bonds to a single metal ion. The most common and most widely used chelating agents are those that coordinate to metal ions through oxygen or nitrogen donor atoms. Phthalic acid coordinates with $Cu^{2+}$ through two oxygen donor atoms, as shown in Figure 6.3.

Table 6.2. Simplified slurry chemistries ("model" slurries) used to examine SiLK and copper CMP mechanisms.

|         | Main Components        | Surfactant | Oxidizer                    | Abrasive                          |
|---------|------------------------|------------|-----------------------------|-----------------------------------|
| Model 1 | DI $H_2O$ Phthalic Acid | None       | None                        | $Al_2O_3$ 0.05 µm (1 wt%)         |
| Model 2 | DI $H_2O$ KH Phthalate  | None       | None                        | $Al_2O_3$ 0.05 µm (1 wt%)         |
| Model 3 | DI $H_2O$ KH Phthalate  | None       | $H_2O_2$ (3.3 vol%)         | $Al_2O_3$ 0.05 µm (1 wt%)         |
| Model 4 | DI $H_2O$ KH Phthalate  | None       | $H_2O_2$ (6.6 vol%)         | $Al_2O_3$ 0.05 µm (1 wt%)         |

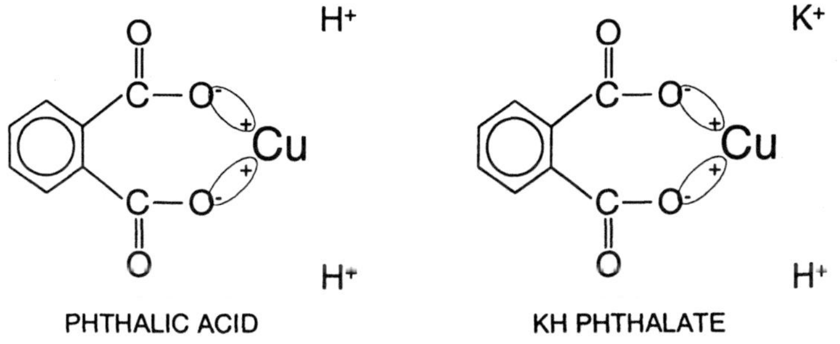

PHTHALIC ACID                    KH PHTHALATE

Figure 6.3. Schematic diagrams showing the chelating action of phthalic acid and potassium hydrogen phthalate.

Copper complexants are good choices for slurry components, as they have been discussed in the literature as necessary components for effective copper removal[6.6]. As illustrated in Figure 6.3, the two organic acids differ only in the fact that KH phthalate has one potassium ion replacing a hydrogen ion. This results in a lower pH level for the phthalic acid slurry, and pH buffering behavior for the KH phthalate slurry.

### 6.2.1 Removal Rate Dependence on Slurry Reactant Concentration

SiLK CMP results with model slurry1 are shown in Figure 6.4. Low concentrations of phthalic acid are effective at achieving SiLK removal; the figure shows the same linear behavior at low concentrations as observed with QCTT1010 in Figure 6.2. However, the SiLK CMP rate decreases sharply from 60 nm/min to 30 nm/min when the phthalic acid concentration exceeds 0.006 M, and remains at this level for higher slurry concentrations. Independent measurements of the acidity of the slurry may suggest a reason for this observed behavior. As mentioned, phthalic acid has two dissociative hydrogens, causing the pH to decrease rapidly as acid concentration is increased. As acid concentration is increased from 0.0015 – 0.006 M, the pH drops from 3.40 to 3.15. At 0.012 M, the pH drops appreciably to 2.55. The highly acidic environment may affect the surface characteristics of the SiLK polymer -- altering either the surface charge or bonding orientation, such that the phthalic acid chemistry is no longer effective at weakening the polymer surface.

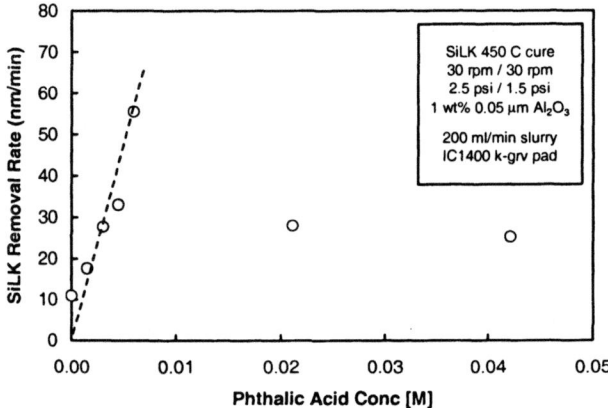

*Figure 6.4.* SiLK removal rate as a function of phthalic acid slurry (slurry model 1) concentration with constant abrasive content

Since phthalic acid enables SiLK CMP reaction and removal at low concentrations, attempts were made to buffer the slurry pH and maintain a value above 3.0 even for higher acid concentrations. KH phthalate selected since it dissociates to form the same chelating chemistry as phthalic acid, but with half of the free hydrogen in solution. CMP results for KH phthalate slurry, model 2, are shown in Figure 6.5. The results show a similar linear increase in removal rate for low concentrations as with the QCTT1010 chemistry. In addition, the buffered pH allows the removal rate to remain high and not decrease as observed in Figure 6.4. The CMP rate plateaus at approximately 175 nm/min as KH phthalate concentration is increased. The buffered pH of these slurries ranges from 5.5 (at 0.0015 M) to 4.25 (at 0.024 M). Thus the pH effect observed with phthalic acid slurry is eliminated, and the SiLK remains susceptible to attack by the reactant.

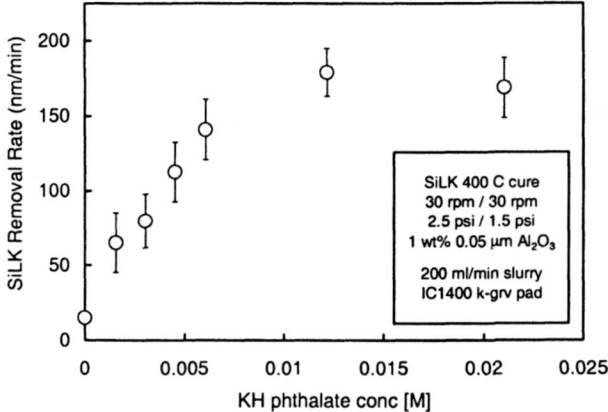

*Figure 6.5.* SiLK removal rate as a function of KH phthalate slurry (slurry model 2) concentration with constant abrasive content.

Due to the noticeable increase in the SiLK removal rate with a relatively small increase in KH phthalate concentration, a test was performed to determine whether or not the chelating properties of the reactant were affecting the abrasive particles. The alumina particles used in the slurries have a net positive surface charge in the acidic KH phthalate media (pH ~ 5.0)[6.7]. There is a concern that this particle charge may be balanced by adsorbed KH phthalate ions, causing particle agglomeration and high removal rate due to enlarged mean particle diameter. To examine this effect, three concentrations of KH phthalate slurry (0.0015, 0.006, and 0.024 M) were measured using a CHDF-2000 particle size measurement system[6.8] from Matec Instruments. Figures 6.6 – 6.9 show the measured slurry particle size distributions for the three slurries.

Two main observations arise from the measurements: (1) the "wet" abrasive particle size differs greatly from the "nominal dry" particle size reported by the vendor; (2) the slurry chemical concentration did not largely affect the slurry particle size in solution. The median particle diameter in each slurry is ~ 400 nm, and high chemical concentrations actually increase the fraction of the particles that have diameter < 100 nm. Thus the chelating slurry chemistry does not cause increased abrasive particle size -- the chemical reaction of the KH phthalate with the SiLK surface is the

126                                                                    *Chapter 6*

mechanism for increased removal rate, not a physical "coagulation" of the slurry particles.

Figures 6.5 – 6.9 show that the KH phthalate slurry chemistry effectively captures the important surface-altering effect of the commercial QCTT1010 chemistry, without increased particle coagulation and scratching damage. Above 0.006 M acid, SiLK removal rate is nearly as high as with the commercial slurry. The phthalic acid appears to react with the SiLK surface, breaking structural bonds, and weakening the surface so that removal by the slurry shear and abrasive is possible. This mechanism is supported by the fact that the phthalic acid slurries have zero static etch rate. Thus the surface-altering reaction is limited to surface coverage, and does not actively etch and dissolve the polymer. This information is used later to propose a phenomenological model (Section 6.3) and a surface kinetics model (Sections 6.4, 6.5) for SiLK CMP.

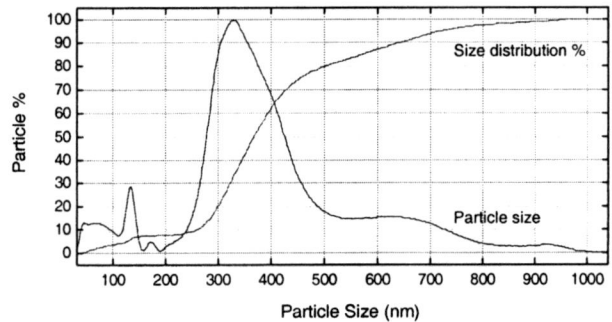

*Figure 6.6.* Particle size distribution of $Al_2O_3$ particles in KH phthalate slurry, with a concentration of 0.0015 M.

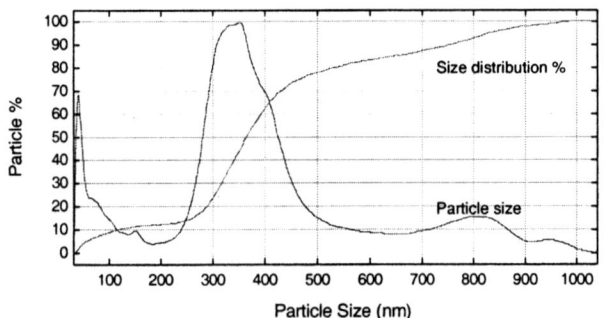

*Figure 6.7.* Particle size distribution of $Al_2O_3$ particles in KH phthalate slurry, with a concentration of 0.006 M.

## LOW-K CMP MODEL BASED ON SURFACE KINETICS

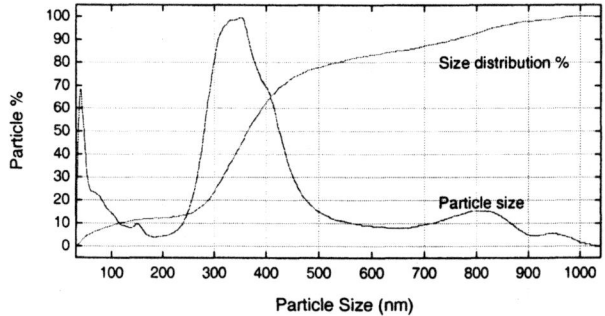

*Figure 6.8.* Particle size distribution of $Al_2O_3$ particles in KH phthalate slurry, with a concentration of 0.024 M.

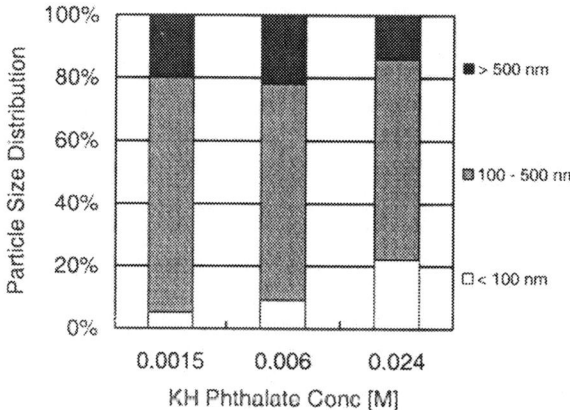

*Figure 6.9.* Bar chart summarizing the particle size distribution of $Al_2O_3$ particles vs KH phthalate concentration in the slurry.

### 6.2.2 KH Phthalate Slurry for Copper CMP Applications

The phthalic acid slurries also remove copper. Model slurries 2, 3, and 4 have been tested for copper CMP removal to determine the relative removal rates of copper and SiLK which enable Cu/low-κ damascene patterning. If a single slurry can remove both the metal and dielectric materials effectively with low scratching damage and varied selectivity, that slurry can act as an LSS during the damascene copper CMP process, resulting in reduced dishing and pattern erosion.

Figure 6.10 shows relative blanket film removal rates for SiLK cured at 400 °C and for sputter deposited, unannealed copper. The data show that added oxidizer is required to obtain high copper removal rate with KH phthalate slurry. Model slurry 2 has ~10 nm/min copper removal rate in the absence of hydrogen peroxide ($H_2O_2$) oxidizer. The same slurry has an order of magnitude higher SiLK removal rate. However, when 3.3 vol% $H_2O_2$ is added to the slurry, the copper removal rate is ~150 nm/min. Copper removal in model slurry 3 results in excellent surface smoothness and low scratch density. It appears that the combination of organic acid and oxidizer results in an effective surface-altering reaction mechanism for copper CMP, similar to results in previous literature[6.6].

The SiLK removal rate is slightly reduced to ~160 nm/min when oxidizer is present in the slurry. It appears that the hydrogen peroxide competes with the phthtalic acid for SiLK surface reactive sites. Peroxide may oxidize a surface site which would otherwise react with the organic acid to create a weakened altered layer. Also of interest is that when the peroxide concentration is doubled, the copper removal rate is reduced to ~ 100 nm/min. A cooperative reaction between slurry oxidizer and complexant may be required to achieve a desirable copper CMP process, in which a specific ratio of oxidizer to complexant enables the highest copper removal rate. The data show tunable copper and SiLK CMP rates. Varying the complexant and oxidizer slurry chemistries can result in the the desired selectivity during damascene patterning or other CMP processes.

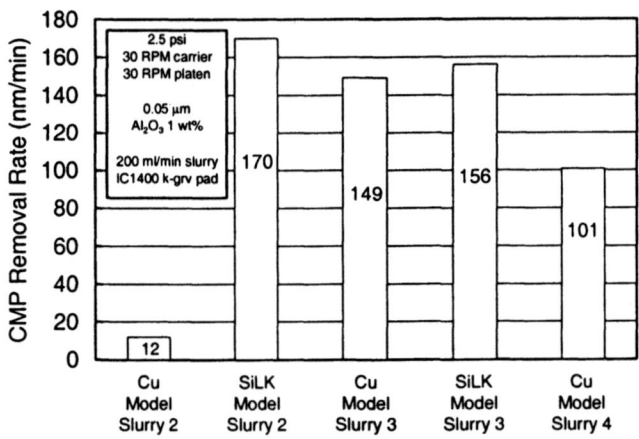

*Figure 6.10.* Removal rate comparison for SiLK and copper using KH phthalate slurry chemistry and hydrogen peroxide oxidizer additive.

### 6.2.3 SiLK Removal Rate Dependence on Velocity

The model slurries have been used to determine the chemical constituents that affect SiLK CMP. Equally important is the physical mechanism for material removal. CMP is a synergistic process where slurry chemistry and physical forces work together. Mechanical removal occurs due to the normal and shear forces developed as a result of the pressure and velocity during CMP. Figure 6.11 shows the SiLK removal rate in slurry model 2 with platen and carrier velocities of 30, 45, and 60 rpm. The results indicate that the velocity affects the value of the asymptote in removal rate. At low reactant concentrations, velocity does not impact the removal rate to a great extent. At reactant concentrations above 0.006 M, the removal rate increases with velocity.

These results illustrate the mechanical aspect involved in SiLK removal. At low reactant concentrations, velocity (shear stress) is less important because the removal rate is dependent on the chemical in the slurry creating the altered layer, which is then removed. Once the reactant concentration is sufficient to adequately form this layer, the physical forces become more important, since the removal of the altered layer becomes the limiting process for CMP. This hypothesis can be tested within the framework of a multi-step surface kinetics model.

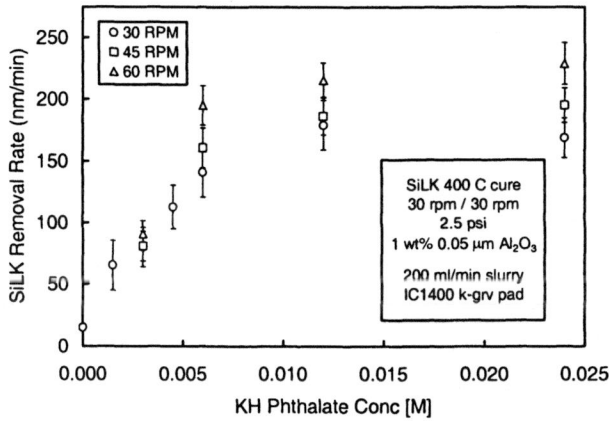

*Figure 6.11.* Removal rate of SiLK using KH phthalate slurry chemistry at different CMP operating velocities (30, 45, 60 rpm carrier and platen).

## 6.3 PHENOMENOLOGICAL MODEL FOR CMP REMOVAL

The previous discussion of removal rate, surface topography, chemical composition, and structural analysis results for polymers and OSG, together with the conceptual framework of the Langmuir-Hinshelwood approach, has been used to develop a physically-based phenomenological model for low-κ CMP[6.9]. The model centers on the concept of an altered layer mechanism that can follow one of two paths, depending on the nature of the slurry chemistry and its effect on the low-κ surface. Figure 6.12 shows the two paths that the mechanism can follow, depending on the interaction of the slurry with the low-κ surface. Figures 6.13 and 6.14 show the complete phenomenological model for both cases, resulting in either high or low material removal rate.

As shown in Figures 6.13 and 6.14, the phenomenological model for low-κ CMP consists of five steps based on fundamental concepts:
1) Diffusion of slurry wetting agent and active reactive species to the wafer surface
2) Adsorption of wetting agent, and contact of reactant with film surface
3) Reaction of wetting agent and reactant with film surface (this step can either result in surface passivation or surface weakening)
4) Shear removal of altered film surface layer by slurry shear forces and abrasive particles
5) Diffusion of CMP product (abraded film) away from the wafer surface, into the bulk slurry

The mechanism has been developed based on a knowledge of BCB, SiLK, and OSG CMP, but can adequately describe the CMP of several of the other low-κ materials in the literature[6.4, 6.10-6.13], and may provide a generic understanding of the CMP process for all low-κ candidates.

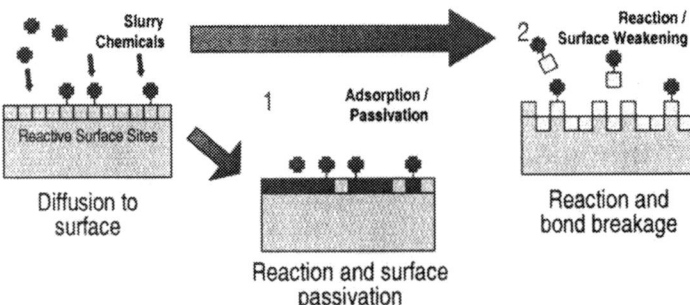

*Figure 6.12.* Phenomenological surface mechanism, showing two cases for low-κ CMP. (1) adsorption and passivation reaction, and (2) cleavage of structural bonds, and surface alteration/weakening.

## 6.3.1 Applicability to the CMP of BCB and SiLK

The model in Figure 6.13 describes the case of low scratching, low removal rate CMP, as observed with BCB in slurries 1 – 4, SiLK in slurries 1,2 and A-C (see Appendix A for slurry descriptions). During the initial stages of CMP the chemicals in the slurry diffuse to the polymer surface. Surfactant and reactant adsorb on the polymer surface and the abrasive, resulting in a monolayer that coats both the dielectric surface and the abrasive particles in solution. The time for diffusion and adsorption determines the latency period depicted in Figures 4.1 and 4.2. The slurry chemistry does not break structural bonds in the polymer surface region, but may form a protective layer by oxidizing the dielectric surface.

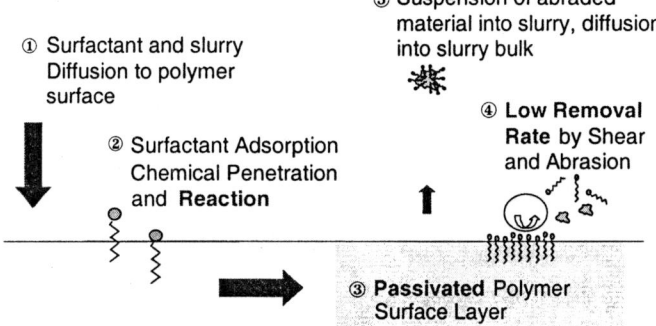

*Figure 6.13.* Phenomenological surface mechanism, showing low removal rate due to adsorption and passivation reaction.

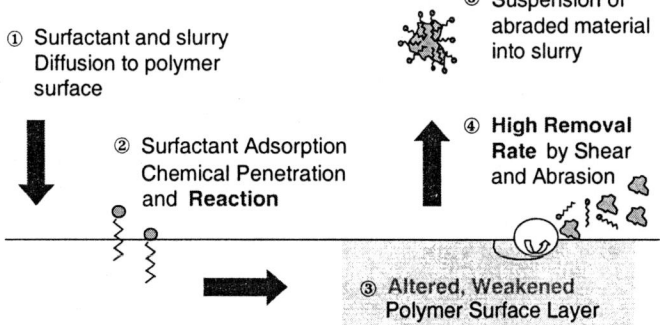

*Figure 6.14.* Phenomenological surface mechanism, showing high removal rate due cleavage of structural bonds, and surface alteration/weakening reaction.

The combined protection of oxidation and surfactant adsorption inhibits fast removal by shear and abrasion. Low removal rate is accompanied by low scratching and low surface damage. Smooth surfaces have been measured for both polymers when surfactant additive is present. The slurry chemical alteration remains at the polymer surface (approximately 10 nm), leaving the bulk unaffected.

Changes in film structure due to different cure time and temperature may be taken into account in this model through a film hardness parameter. Shorter cure times and lower cure temperatures would result in a lower hardness parameter, which would increase the removal rate accordingly. Observations suggest that a decrease in removal rate occurs due to increased film crosslinking at elevated time and temperature. Since mechanical film removal dominates, the softer films have the same surfactant adsorption latency, but are removed more readily due to shear and abrasion.

The model in Figure 6.14 describes the case of low scratching, high removal rate CMP observed with SiLK in slurries 3-4 and model 2-4. The slurry wets the SiLK surface and slurry chemicals react rapidly with the exposed polymer. The speed of the chemical reaction is indicated by the lack of a measurable latency period as depicted in Figure 4.2. Slurry chemicals break structural polymer bonds and form a weakened surface layer. Evidence of an oxidized SiLK surface layer is apparent in the surface XPS results discussed in Chapter 4.

The altered layer is easily removed by the shear stress and abrasion of CMP, resulting in high removal rates. In addition to the oxidizing attack of the slurry chemicals, surfactant adsorbs on the polymer and abrasive and protects the polymer surface from scratches, as reported in the AFM roughness data. A synergy is apparent between the chemical attack of the polymer film, which breaks structural bonds, and the mechanical forces at the film surface due to the pressure and shear of the CMP process. The continuous synergistic action of CMP creates an altered chemical layer that is rapidly removed, while the underlying polymer material remains both physically and chemically unaffected.

Changes in film structure due to different cure conditions may be taken into account in this model through parameters for film hardness and number of chemically reactive sites. Observations suggest that an increase in the removal rate occurs due to reduced film crosslinking and increased number of chemically reactive sites in the film structure. Since both mechanical and chemical interactions contribute to the synergistic CMP removal, the softer and more chemically susceptible films are removed by CMP at a higher rate.

## 6.3.2 Applicability to the CMP of Organosilicate Glasses

Experimental results for film removal rate, surface roughness, and surface and bulk chemical change can also be used to develop a phenomenological understanding of the CMP removal of PECVD OSG materials. The hypothesized mechanism combines those for silicon dioxide and organic polymer thin films, and takes into account the differences observed between the three OSG derivatives and SiLK to understand better the removal mechanisms of these new materials. Several experimental observations correspond well to the model proposed in Figure 6.14:

1) OSG CMP removal rates are much higher than organic SiLK CMP rates in slurries designed for $SiO_2$ CMP.
2) OSG CMP removal rates increase when films absorb water from the ambient.
3) OSG CMP removal results in smooth, damage-free surfaces at high slurry pH.
4) OSG CMP removal rates increase with slurry pH and decrease with film carbon content.
5) The post-CMP surface has the same chemical composition as the OSG bulk.

Observations (1) - (3) illustrate the importance of the surface-weakening chemical reaction between the slurry and the OSG surface, as observed in SiLK CMP[6.9]. For rapid CMP removal, the slurry chemicals must react with the OSG surface to form a weakened layer, which can be removed by the abrasive and shear forces within the CMP slurry. Observation (1) shows that the high pH slurries designed for $SiO_2$ CMP alter and weaken the OSG surface, but do not react with the organic polymer (SiLK) surface.

It is likely that a selective hydroxyl-based chemistry attacks Si-O bonds much more readily than Si-C bonds, and that this is the key component of the OSG CMP process. This mechanism is supported by the fact that water penetration into the films increases the OSG removal rate. Together, observations (1) – (3) suggest that successful CMP removal of OSG depends on the scission of Si-O structural bonds at the film surface, due to the chemical hydrolysis reaction reported by Cook for the CMP of $SiO_2$[6.14]:

$$\equiv Si-O-Si \equiv \ + \ OH^- \ \xleftrightarrow{k_1} \ \equiv Si-O^- \ + \ \equiv Si-OH \quad (6.1)$$

This reaction is enhanced by absorbed water, which also penetrates the OSG structure and hydrolyzes Si-O bonds by the reaction[6.14]:

$$\equiv Si-O-Si \equiv \ + \ H_2O \ \xleftrightarrow{k_2} \ 2 \equiv Si-OH \quad (3.1)$$

It does not appear that the slurry and ambient water react with the Si-C or C-H bonds. The reactions in Equations 6.1 and 3.1 are suppressed by the carbon content of the film, due to the increased film hydrophobicity and reduced number of reactive surface Si-O sites. The slurry chemistry does not attack the Si-C or C-H bonds in the films, but is able to break down the oxide-like structure of the films. Since the OSG films contain substantial carbon, their removal rates in these slurries are typically lower than the removal rate of plasma deposited $SiO_2$. The slurry chemistry reacts with the Si-O bonds, forming a weakened dielectric surface layer of fragmented OSG. The shear and abrasion of the slurry remove this weakened layer. Material is then removed in groups of silicon with attached carbon groups, as shown in Figure 6.15. The Si and C are removed together in small clusters rather than separately as $Si(OH)_4$ or $CH_4$. As per observation (5), the carbon in the film is removed at the same rate as the inorganic content since the methyl groups remain bound to the silicon atoms in the chain.

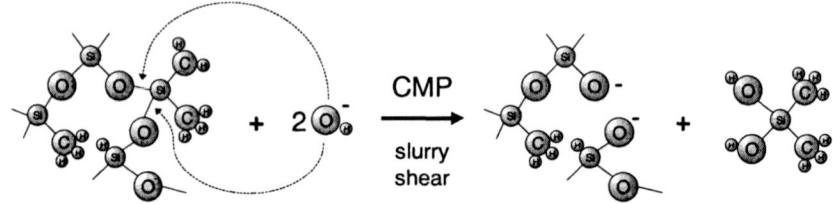

*Figure 6.15.* Proposed mechanism for OSG CMP removal by slurry chemistry and shear.

The proposed mechanism results in constant Si, O, and C atomic percentages in the OSG surface before and after CMP. CMP removal occurs by an altered-surface removal mechanism which results in a smooth surface due to the fact that the chemically altered surface layer is continuously removed by the CMP slurry shear.

## 6.4 FIVE STEP REMOVAL MODEL FOR SILK CMP USING MODIFIED LANGMUIR-HINSHELWOOD KINETICS

The phenomenological models described previously provide a framework on which to build a more fundamental mechanism for low-κ CMP. A complex, multi-step mechanism can be formulated to include each of the steps described in the phenomenological models. The basis for such a surface reaction mechanism is Langmuir-Hinshelwood (L-H) approach to multi-step surface kinetics using heterogeneous catalysis [6.15]. This is used to pose the CMP problem in terms of a five-step surface mechanism that can be represented mathematically. The surface mechanism is then used as a the wafer-site boundary condition for mass transport equations that can be solved when coupled with the fluid mechanics equations that govern the slurry flow between the pad and wafer. The L-H method allows one to assume a particular rate-limiting step, and generate a solution for that case to test to experimental data. This procedure allows examination of one mechanistic step at a time. Comparison with experimental results provides fundamental insight into the controlling mechanisms for the SiLK CMP process.

### 6.4.1 Five Step Surface Mechanism

The five-step surface mechanism is a mathematical representation of the phenomenological progression shown in Figures 6.13 and 6.14. The five steps include:

(i)  mass transport of reactant from the bulk slurry to the slurry/wafer interface
(ii)  adsorption of reactant to available SiLK polymer surface sites
(iii)  reaction between adsorbed reactant and the SiLK polymer surface site to which it is attached
(iv)  shear-enhanced desorption of weakened altered polymer surface layer
(v)  mass transport of polymer product from the slurry wafer interface to the bulk slurry

This mechanism is similar to heterogeneous catalysis; however, it must be modified to account for the fact that the catalyst surface, in this case, the polymer itself, is consumed rather than conserved. Figure 6.16 shows a schematic depiction of the multi-step surface mechanism, showing forward and reverse reactions and rate constants for adsorption, surface reaction, and shear-enhanced desorption. In step (iv), the shear-enhanced desorption, polymer product is removed, leaving behind a new polymer surface site for reaction to take place, and the process to repeat. Conservation of sites in this manner is crucial to representing the model using this heterogeneous catalysis scheme.

Step (i) is mass transport of the slurry reactant from the bulk slurry to the slurry/wafer interface, which may be described as:

$$A_{active}|_{slurry,L} \xleftrightarrow{k_{m1}} A_{active}|_{interface,L} \quad (6.2)$$

with the corresponding rate for mass transfer, based on a mass transfer coefficient for the reactant, $k_{m1}$:

$$r_{m1} = k_{m1}\left(A_{active}|_{slurry,L} - A_{active}|_{interface,L}\right) \quad (6.3)$$

Each species concentration in the slurry/wafer interface ($A$) has units of mol/m$^3$.

Step (ii) is the adsorption of the slurry reactant from the interface layer to a free polymer surface site:

$$A_{active}|_{interface,L} + \underline{P} \xleftrightarrow{k_{ads}} A_{active}|_S \quad (6.4)$$

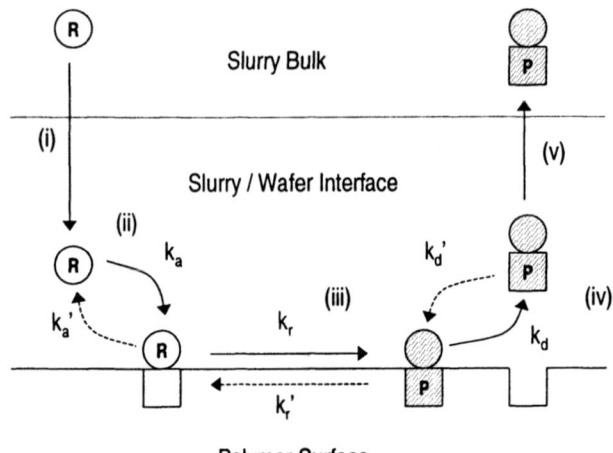

*Figure 6.16.* Schematic representation of five-step modified Langmuir-Hinshelwood surface kinetics model for SiLK CMP, based on heterogeneous catalysis. Note that the polymer surface is consumed rather than conserved.

$$r_{ads} = k_{ads} \underline{P} A_{active}\big|_{interface, L} - k'_{ads} A_{active}\big|_S \quad (6.5)$$

where $k_{ads}$ and $k'_{ads}$ are the forward and reverse rate constants for adsorption, and $\underline{P}$ denotes an available polymer surface site.

Step (iii) is the surface reaction between the adsorbed reactant species and the polymer surface site:

$$A_{active}\big|_S \xleftrightarrow{k_{rxn}} A_{altered}\big|_S \quad (6.6)$$

$$r_{rxn} = k_{rxn} A_{active}\big|_S - k'_{rxn} A_{altered}\big|_S \quad (6.7)$$

where $k_{rxn}$ and $k'_{rxn}$ are the forward and reverse rate constants for surface reaction.

Step (iv) is the shear-enhanced desorption of the altered layer, where the reacted polymer surface is removed in pieces due to the shear stress and abrasion during CMP:

$$A_{altered}\big|_S \xleftrightarrow{k_{des}} A_{altered}\big|_{interface, L} + \underline{P} \quad (6.8)$$

$$r_{des} = k_{des} A_{altered}\big|_S - k'_{des} \underline{P} A_{altered}\big|_{interface, L} \quad (6.9)$$

where $k_{des}$ and $k'_{des}$ are the forward and reverse rate constants for shear-enhanced desorption of the polymer surface into the slurry. We propose that $k_{des}$ is a function of the shear stress ($\tau_w$), abrasive concentration ($c_a$), abrasive size ($d_p$), and temperature ($T$) such that $k_{des} = k_{des}(\tau_w, c_a, d_p, T)$. There is question as to the feasibility of allowing a reverse reaction based on $k'_{des}$, because the process in which a physically removed piece of reacted polymer will bind itself back to the polymer surface is unlikely. In this formulation, we have included the reverse reaction for completeness.

Step (v) is mass transport of the abraded polymer product from the slurry/wafer interface to the bulk slurry, which may be described as:

$$A_{altered}|_{interface,L} \xleftrightarrow{k_{m2}} A_{altered}|_{slurry,L} \tag{6.10}$$

with the corresponding rate for mass transfer, based on a mass transfer coefficient for the polymer product, $k_{m2}$ of:

$$r_{m2} = k_{m2}\left(A_{altered}|_{interface,L} - A_{altered}|_{slurry,L}\right) \tag{6.11}$$

Equations 6.2 - 6.11 mathematically describe the surface mechanism and interaction of slurry reactant with SiLK polymer. The combination of these equations determines the boundary condition at the wafer surface. If one step in this process is much slower than the others, it will control the overall rate. Therefore, the equations may be solved in groups, to generate a boundary condition for the flux of the reactant and product species, based on one surface mechanism step being slow, or rate determining. The other steps are assumed to be fast, or in equilibrium. This eliminates forward and reverse reaction rate constants, replacing them with an equilibrium constant for adsorption, reaction, or desorption.

For example, if we choose the case when surface reaction is slow, and thus rate determining (i.e. the rate of the surface reaction determines $RR$, the rate of CMP removal), the set of equations becomes:

(rxn RDS) $\quad r_{rxn} = k_{rxn} A_{active}|_S - k'_{rxn} A_{altered}|_S = RR \tag{6.12}$

(ads equilibrium) $\quad A_{active}|_S = K_{ads} \underline{P} A_{active}|_{interface,L} \tag{6.13}$

(des equilibrium) $\quad A_{altered}|_S = \dfrac{1}{K_{des}} \underline{P} A_{altered}|_{interface,L} \tag{6.14}$

where $K_{ads}$ and $K_{des}$ are equilibrium constants for adsorption of the reactant to the polymer surface and desorption of the polymer product from the surface, respectively.

Equations 6.12, 6.13, and 6.14 may be combined using a surface site balance requirement which states that the total number of polymer surface sites, $S$, must equal the available polymer surface sites, $\underline{P}$, plus the sites with adsorbed reactant, $A_{active}|_S$ plus the chemically altered reacted sites, $A_{altered}|_S$:

$$S = \underline{P} + A_{active}|_S + A_{altered}|_S \tag{6.15}$$

The result of the combination of equations, for the case of surface reaction rate determining (or controlling), is:
(rxn RDS)

$$r_{rxn} = k_{rxn}\underline{P}\left[\frac{K_{ads}A_{active}|_{interface,L} - \dfrac{1}{K_{rxn}K_{des}}A_{altered}|_{interface,L}}{1 + K_{ads}A_{active}|_{interface,L} + \dfrac{1}{K_{des}}A_{altered}|_{interface,L}}\right] = RR$$

(6.16)

where $K_{rxn}$ is the equilibrium constant for reaction between the active chemical component in the slurry and the SiLK polymer film.

Similar solutions can be obtained for the cases where adsorption and desorption are chosen to be slow, or rate determining:

(ads RDS)

$$r_{ads} = k_{ads}\underline{P}\left[\frac{A_{active}|_{interface,L} - \dfrac{1}{K_{ads}K_{rxn}K_{des}}A_{altered}|_{interface,L}}{1 + \left(\dfrac{1}{K_{rxn}K_{des}} + \dfrac{1}{K_{des}}\right)A_{altered}|_{interface,L}}\right] = RR$$

(6.17)

(des RDS)

$$r_{des} = k_{des}\underline{P}\left[\frac{K_{rxn}K_{ads}A_{active}|_{interface,L} - \dfrac{1}{K_{des}}A_{altered}|_{interface,L}}{1 + K_{ads}(1 + K_{rxn})A_{active}|_{interface,L}}\right] = RR$$

(6.18)

and Equations 6.16 - 6.18 can be placed in the generic form:

$$RR = K_1\left[\frac{K_2 A_{active}|_{interface,L} - K_3 A_{altered}|_{interface,L}}{1 + K_4 A_{active}|_{interface,L} + K_5 A_{altered}|_{interface,L}}\right] \quad (6.19)$$

where $K_1$ is a constant that is proportional to the forward rate constant of the rate-determining mechanistic step (i.e. $K_1 = k_{ads}\underline{P}$, $k_{rxn}\underline{P}$, or $k_{des}\underline{P}$), and $K_2$, $K_3$, $K_4$, $K_5$ are constants which are functions of the equilibrium constants $K_{ads}$, $K_{rxn}$, and $K_{des}$.

In summary, simplification using L-H kinetics generates a flux condition at the wafer surface which is based on a rate-determining (or controlling) surface mechanism. The removal rate for each case can be modeled and the results tested against the experimental data. The inclusion of the flux boundary condition into a complete framework of model equations for the CMP process is not trivial, as discussed in the next section.

## 6.4.2 Implementation into 3-D Fluid Mechanics and Mass Transport Models

A three-dimensional fluid-mechanics and mass-transport CMP model has been developed by Thakurta et al.[6.2], who extended the work of Sundararajan et al.[6.1] as described previously in section 3.4.2. This model is the framework with which to solve the complex multi-step CMP surface kinetics equations.

Briefly, the model solves the three-dimensional Navier-Stokes equation for laminar, lubricating slurry flow within the geometry created by the rotating wafer and pad. The solution generates a velocity and pressure profile for every point within the slurry film between the wafer and pad. The information on the velocity distribution is then used as an input to a finite-element mass-transport code which solves the three-dimensional diffusion equations for both the reactant and product species:

$$u \frac{\partial A_{active}|_{slurry,L}}{\partial x} + v \frac{\partial A_{active}|_{slurry,L}}{\partial y} + w \frac{\partial A_{active}|_{slurry,L}}{\partial z} = D_{active} \frac{\partial^2 A_{active}|_{slurry,L}}{\partial z^2} \quad (6.20)$$

$$u \frac{\partial A_{altered}|_{slurry,L}}{\partial x} + v \frac{\partial A_{altered}|_{slurry,L}}{\partial y} + w \frac{\partial A_{altered}|_{slurry,L}}{\partial z} = D_{altered} \frac{\partial^2 A_{altered}|_{slurry,L}}{\partial z^2} \quad (6.21)$$

Equations 6.20 and 6.21 require boundary conditions (B.C.) to obtain solutions for $A_{active}|_{slurry,L}$ and $A_{altered}|_{slurry,L}$. The system geometry is shown in two dimensions in Figure 6.17. The equations require two B.C. for $A_{active}|_{slurry,L}$ and $A_{altered}|_{slurry,L}$ in the z-direction, and one B.C. each in the x, y directions.

One B.C. in x and y for all z is satisfied by stating the following slurry inlet condition:

$$A_{active}|_{slurry,L}|_{inlet} = constant \; (specified \; by \; user) \quad (6.22)$$

$$A_{altered}|_{slurry,L}|_{inlet} = 0 \quad (6.23)$$

which assumes that fresh slurry always enters the wafer domain, with no product present. The slurry reactant concentration is specified by known experimental data.

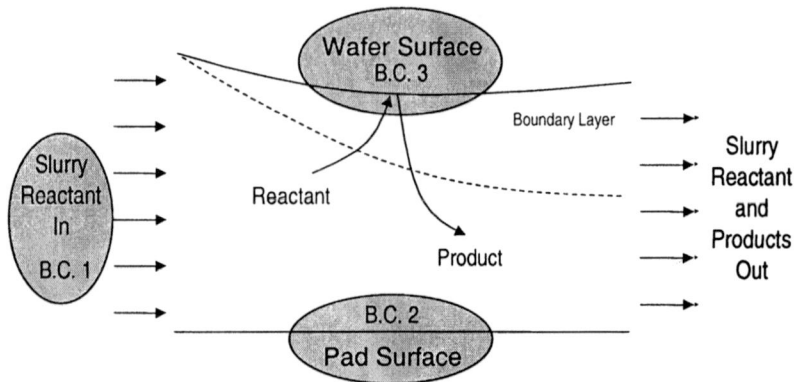

*Figure 6.17.* Side view schematic of the solution domain for Thakurta's 3-D fluid-mechanics and mass-transport based CMP model, showing the boundary conditions selected to enable solution. From ref[6.2].

The next B.C. states that the pad surface is impenetrable to slurry reactant and product:

$$\frac{\partial A_{active}|_{slurry,L}}{\partial z}\bigg|_{pad,z=0} = 0 \quad (6.24)$$

$$\frac{\partial A_{altered}|_{slurry,L}}{\partial z}\bigg|_{pad,z=0} = 0 \quad (6.25)$$

And the final B.C. in z is the flux condition at the wafer surface, which has the form of the rate-determining step equations, and follows from Equation 6.19. Note the difference in signs in Equations 6.26 and 6.27:

$$-D_{active}\frac{\partial A_{active}|_{slurry,L}}{\partial z}\bigg|_{wafer,z=h} = K_1\left[\frac{K_2 A_{active}|_{interface,L} - K_3 A_{altered}|_{interface,L}}{1 + K_4 A_{active}|_{interface,L} + K_5 A_{altered}|_{interface,L}}\right] \quad (6.26)$$

$$D_{altered}\frac{\partial A_{altered}|_{slurry,L}}{\partial z}\bigg|_{wafer,z=h} = K_1\left[\frac{K_2 A_{active}|_{interface,L} - K_3 A_{altered}|_{interface,L}}{1 + K_4 A_{active}|_{interface,L} + K_5 A_{altered}|_{interface,L}}\right] \quad (6.27)$$

When Equations 6.20 and 6.21 are simultaneously solved with the coupled boundary conditions of Equations 6.22 - 6.27, the result is a concentration profile for $A_{active}|_{slurry,L}$ and $A_{altered}|_{slurry,L}$ at all points within the slurry film. This information is used to calculate a flux of $A_{active}|_{slurry,L}$ to the wafer surface and $A_{altered}|_{slurry, L}$ away from the wafer surface, which is related to the CMP removal rate:

$$RR = -\frac{1}{n}\frac{MW_{altered}}{\rho_{altered}} D_{active} \frac{\partial A_{active}|_{interface, L}}{\partial z} \quad (6.28)$$

where
- $RR$ is the predicted SiLK removal rate
- $n$ is a stoichiometric constant that specifies the number of reactant molecules required to alter one section of the SiLK polymer structure
- $MW_{altered}$ is the molecular weight of one altered section of the SiLK polymer structure (the product that desorbs from the wafer surface)
- $\rho_{altered}$ is the density of the altered polymer product
- $D_{active}$ is the diffusivity of the active reactant component in the slurry

Thus, solution of the fluid mechanics and mass transport equations with the coupled boundary conditions generates a prediction for SiLK polymer removal rate, based on the inlet slurry reactant concentration and several adjustable constants.

### 6.4.3 Results

The model equations listed above have been used to calculate SiLK CMP removal rate for a variety of slurry reactant concentrations, to compare with experimental data. For a given reactant concentration, the model calculates the concentration profiles of reactant and product under the wafer surface, as well as the removal rate of SiLK at different locations on the wafer.

Solution of the model equations requires values for the many constants in Equation 6.26 and 6.27. Table 6.3 lists the rate constants that have been selected for the model equations. These constants are selected from the literature for adsorption, reaction, and desorption of organic species, and provide a reasonable first estimate for our model calculations.

The adsorption equilibrium constant, $K_{ads}$, which represents the adsorption equilibrium of a small organic acid molecule to a hydrophobic organic polymer surface, is estimated at a value of $K_{ads} = 1.0 \times 10^4$ m$^3$/mol. This value is taken from the measured adsorption of napthalene sulfonate (a double-benzene small organic molecule) to methylated/alkylated hydrophobic silica surfaces. For physical adsorption of hydrophobic species in liquid systems, $K_{ads}$ may range from $1.0 \times 10^2$ for small molecules to $1.0 \times 10^5$ m$^3$/mol for large molecules[6.16].

Table 6.3. Equilibrium constants used in multi-step L-H kinetics model for SiLK CMP

| Surface Step | Value | Comments | Author Reference |
| --- | --- | --- | --- |
| Adsorption ($K_{ads}$) | $1.0 \times 10^4$ M$^{-1}$ | Estimated from literature for adsorption of (1-anilino-8-napthalene sulfonate) on methylated and alkylated silica surfaces | Harris [6.16],[6.17] |
| Reaction ($K_{rxn}$) | $1.0 \times 10^0$ | Estimated from literature for bond cleavage in benzyl phenyl ethers at ~ 100 °C | Penn [6.18] |
| Desorption ($K_{des}$) | $1.0 \times 10^4$ M | Estimated from literature for desorption of rhodamine6G+ cationic dye from alkylated silica surface | Harris [6.19] |

The reaction equilibrium constant, $K_{rxn}$, which represents the equilibrium reaction for bond cleavage in the organic polymer surface (either C-O or C-C bonds), is estimated at a value of $K_{rxn} = 1.0 \times 10^0$ (unitless). This value is taken from the measured reaction rate of the bond cleavage of organic phenyl esters (benzene-O-benzene) at approximately 100 °C. For the temperatures and pressures observed during CMP, $K_{rxn}$ may range from $1.0 \times 10^{-1}$ for inert material/slurry combinations at room temperature to $1.0 \times 10^4$ for reactive material/slurry combinations[6.18].

The desorption equilibrium constant, $K_{des}$, which represents the desorption equilibrium of a large organic molecule from a hydrophobic organic polymer surface, is estimated at a value of $K_{des} = 1.0 \times 10^4$ mol/m$^3$. This value is taken from the measured desorption of Rhodamine6G+, a large-chain organic dye molecule, from an alkylated silica surface. This constant represents strictly physico-chemical desorption, and does not take into account any shear-enhancement of species desorption. For physical adsorption of hydrophobic species in liquid systems, $K_{des}$ may range from $1.0 \times 10^2$ for large molecules to $1.0 \times 10^5$ m$^3$/mol for small molecules[6.19].

Using the equilibrium constants listed above, the adjustable parameters used to obtain a model fit are the stoichiometric coefficient, $n$, in Equation 6.28 and the constant $K_1$ in Equations 6.26 and 6.27. The best model fits are obtained for $n = 12$, which allows $RR$ predictions that match experimental values for the entire concentration range of $A_{active}$. A stoichiometric constant with a value > 1 is expected in the case of a single reactant molecule reacting with a highly crosslinked polymer -- several reactant molecules are required to alter one polymer unit. With $n = 12$, the model equations have been solved and $K_1$ has been adjusted to calculate the SiLK CMP removal rate for a variety of slurry reactant concentrations to compare with experimental data.

For a given reactant concentration, the model is used to calculate the concentration profiles of reactant and product under the wafer surface, as well as the removal rate of SiLK at different locations on the wafer. Figures

6.18(a-e) show the model calculations for the case of desorption RDS, $A_{active}|_{inlet}$ = 0.024 M, $V_{carrier}$ = 30 rpm, $V_{platen}$ = 30 rpm, and $P$ = 2.5 psi. Figures 6.18(a-d) show slices of the lubricating slurry layer at four heights within the slurry film between the wafer and pad surfaces. 6.18(a) shows the slurry concentration contours at the wafer surface, 6.18(b) shows the contours at 75% of the slurry film height, 6.18(c) is midway between the pad and wafer, and 6.18(d) is located near the pad surface, at 25% of the slurry film height. Figures 6.18(a-e) show that the slurry bulk is only slightly depleted in the KH phthalate reactant due to the surface reaction at the wafer surface. Thus, for $A_{active}|_{inlet}$ = 0.024 M the SiLK CMP process is reactant-abundant as observed by the plateau in experimentally observed removal rate.

Figure 6.19(a) shows contours for the SiLK removal rate across the wafer surface, and Figure 6.19(b) shows the average SiLK removal rate predicted by the model as a function of position along the wafer radius. That the model predicts slightly higher removal rate near the wafer edge (radius = 70 millimeters) due to the increased reactant concentration in the incoming slurry. The slurry enters fresh, at its highest reactant concentration, near the wafer edge, and becomes depleted as it passes under the wafer towards the wafer center. However, the difference in removal rate is very small due to the abundance of the chemical reactant in the slurry. In the case of reactant-depleted CMP ($A_{active}|_{inlet}$ = 0.0015 M) the model predicts a much larger variation in the removal rate between the wafer edge and the wafer center, due to depletion of the reactant at the wafer edge.

Model calculations such as those shown in Figures 6.20 and 6.21 are used to obtain best fits to the experimental SiLK RR data. Comparison of the model predictions for different selected rate determining steps allows comparison of different controlling mechanisms for SiLK removal. We have performed calculations for $n$ = 12 by varying $K_1$ in each case of adsorption, reaction, and desorption RDS, to obtain the best overall fit to the removal rate data. Calculations are plotted and compared for $V_{carrier}$ = 30 rpm, $V_{platen}$ = 30 rpm, and $P$ = 2.5 psi, as shown in Figure 6.20.

The adsorption rate-limiting case provides the worst fit to the data, since the predicted removal rate increases nearly linearly with reactant concentration. This behavior is expected based on L-H kinetics. Adsorption is rarely the RDS for surface heterogeneous catalysis since physical adsorption to a solid surface occurs rapidly on the time scale of surface reaction and desorption in both the gas and liquid phases.

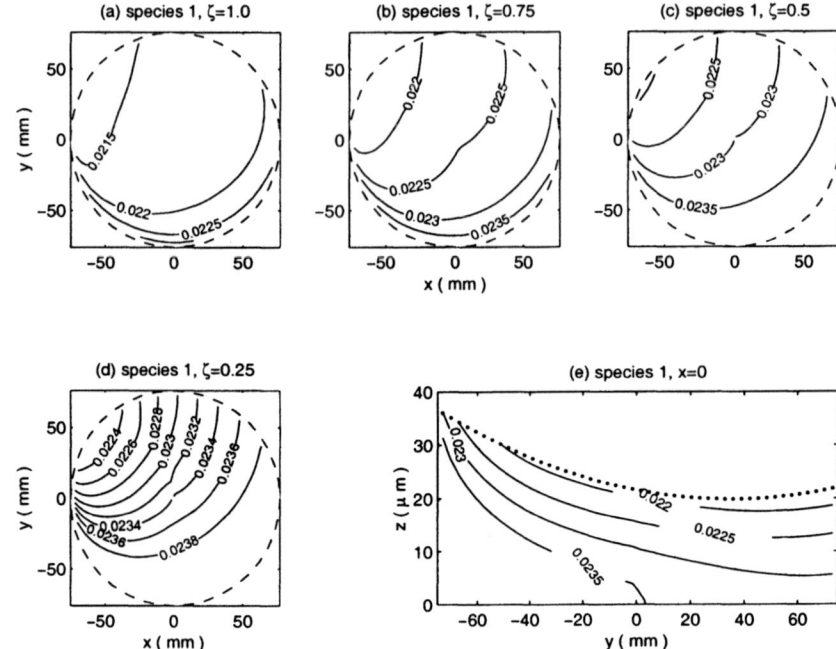

*Figure 6.18.* Concentration contours of Aactive during SiLK CMP, desorption RDS, $A_{active|inlet}$ = 0.024 M, $V_{carrier}$ = 30 rpm, $V_{platen}$ = 30 rpm, P = 2.5 psi. (a-d) show slices of the slurry film at 100%, 75%, 50%, and 25% of the slurry film height between the wafer and pad surfaces. (e) shows concentration profiles for a cross section along the wafer diameter.

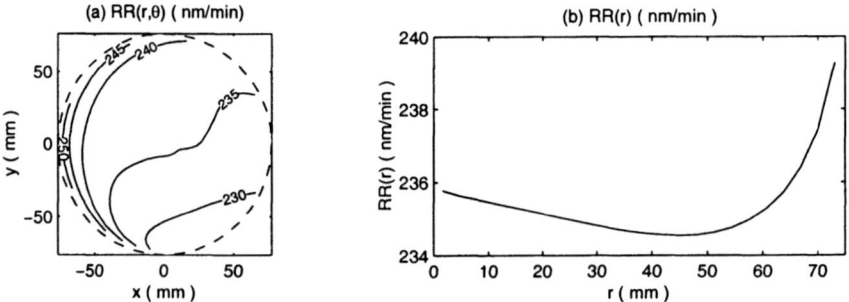

*Figure 6.19.* Contours of SiLK CMP removal rate for desorption RDS, $A_{active|inlet}$ = 0.024 M, $V_{carrier}$ = 30 rpm, $V_{platen}$ = 30 rpm, P = 2.5 psi. (a) shows variation in removal rate across the wafer surface, and (b) shows the variation in average removal rate across the wafer radius.

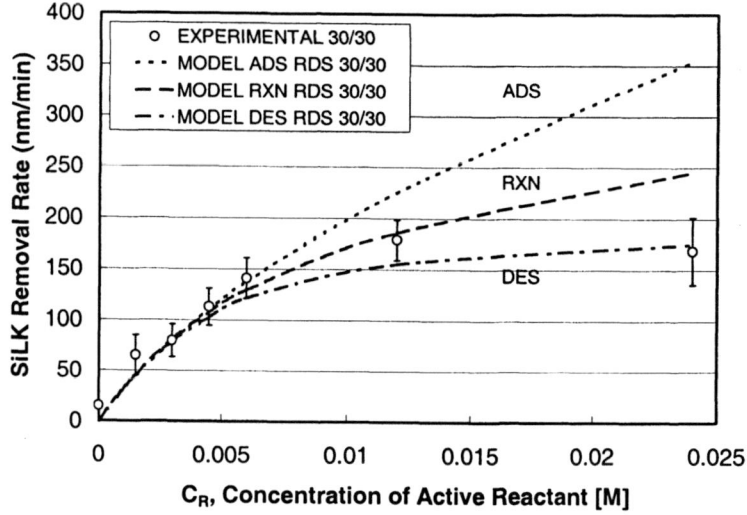

*Figure 6.20.* Model comparison to experimental data for SiLK CMP removal rate as a function of slurry active reactant component. Data fits are provided for adsorption, surface reaction, and shear-enhanced desorption rate-determining cases

The case of the surface reaction rate-limiting step provides a better overall fit to the data, and fits the data at low reactant concentrations very well. However, the model does not predict the "plateau" in removal rate at higher reactant concentrations that is observed in experiments. This may represent a transition from reaction rate dependence (at low reactant concentrations) to desorption rate dependence (at high reactant concentrations). The shear-enhanced desorption rate-limiting case provides the best overall fit to the data, including the rapid rise in removal rate at low reactant concentration and the leveling of removal rate at high reactant concentration. A combination of mechanisms may also explain the statistically significant drop in removal rate that is observed at the highest reactant concentration. When $A_{active}|_{inlet} = 0.024$ M, the removal rate drops slightly, in a manner observed previously with copper and FLARE low-κ polymer[6.4]. Although the desorption RDS data provides a good fit to the experimental data, a combined reaction and desorption mechanism may capture the second order effects observed with removal rate versus chemical concentration in the real CMP system.

The concept of a transition between controlling mechanisms makes physical sense, due to the nature of the CMP process. The slurry active component reacts with the SiLK surface to create an altered layer which is then removed mechanically. At low slurry concentrations, the amount of reactant may be insufficient to create a completely reacted monolayer, covering the wafer surface. Increasing reactant concentration directly affects the amount of coverage of the "altered layer", so at low concentrations, the amount of reactant is vital to the removal rate. At high reactant concentrations, a complete layer of altered material exists on the wafer surface, and the shear removal of this altered layer determines the CMP removal rate. Thus, in the high-concentration regime, the reactant concentration does not linearly impact the removal rate. Since the solution for desorption RDS fits both regimes of the data very well, model solutions have been tested for various CMP conditions using the desorption RDS model.

Model calculations were used to test the impact of the pad and wafer velocity on the mass transport resistance to the CMP process. The fluid mechanics flow field was calculated for three different slurry velocities (30, 45, and 60 rpm), to evaluate the effect of velocity on removal rate when $K_1$ is held constant at the value of $1.7 \times 10^{-4}$ mol/m$^2$ s used to fit the desorption case in Figure 6.20. Figure 6.21 shows how the increased platen/carrier velocity impacts the SiLK removal rate measured experimentally.

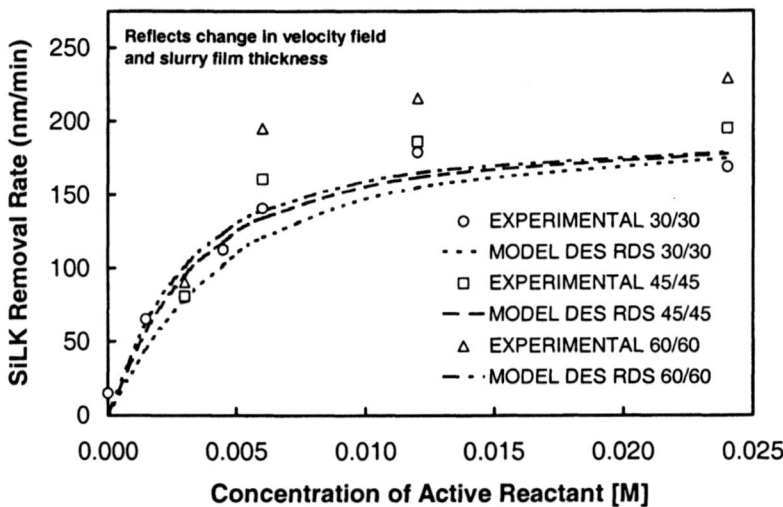

*Figure 6.21.* Model predictions showing variation in SiLK CMP removal rate with CMP velocity, for the case of shear-enhanced desorption RDS.

The model result predicts an increase in the SiLK removal rate with the velocity, but not the same dependence observed from experiments. The velocity has the largest impact on the model predictions at low reactant concentrations (reactant-depleted regime), where mass transport is most likely to be of importance. However, it can be seen from Figure 6.21 that the mass-transport enhancement provided by the velocity increase (the corresponding enhanced diffusivity of reactant and product) is not solely responsible for the overall increase in CMP rate. Velocity has an additional effect on the CMP process through the chemical/physical synergism described in section 3.2.1.

The relative importance of mass transport in a system can be determined by calculating the Sherwood Number, $N_{Sh}$, for the specific system geometry and species concentrations. $N_{Sh}$ for transport of the reactant to the wafer surface is defined as:

$$N_{Sh} = \frac{k_{ml} h}{D_{active}} \qquad (6.29)$$

where
- $k_{ml}$     is the mass transport coefficient for the active component (KH phthtalate)
- $h$     is the slurry height, or the gap distance between the wafer and pad
- $D_{active}$     is the diffusivity of the active reactant component in the slurry

$N_{Sh}$ is the dimensionless ratio of the total mass transport in the system to the diffusive mass transport. A value of $N_{sh} \leq 1$ is characteristic of a mass-transport limited system. For our CMP process, the mass transport coefficient is calculated from Equation 6.2 by equating $r_l$ to the flux of $A_{active}$ to the wafer surface, which is calculated from the removal rate. Values for $A_{active}|_{slurry,L}$ and $A_{active}|_{interface,L}$ are average values of the reactant concentration in the slurry bulk and at the wafer surface, calculated from the model output (i.e. Figure 6.18). The slurry height, $h$, is calculated by the lubrication model for a given pressure and velocity, and the diffusivity of the reactant, $D_{active}$, is taken to be $10^{-5}$ cm$^2$/s. For a wide range of CMP velocities, $N_{Sh} \sim 6.0 - 7.0$, supporting our findings that the diffusive mass transport plays a secondary role under the conditions used here during the SiLK CMP process.

Model results using Equation 6.16 for the case of shear-enhanced desorption have been re-calculated for the three velocities shown in Figure 6.21 by changing $K_l$ to fit the experimental data, with the result shown in Figure 6.22. This illustrates the dependence of the rate of shear-enhanced desorption, $k_{des}$, on the physical forces (shear, abrasion) that exist during CMP. Velocity enhances the shear rate, which has an effect on $k_{des}$. This is reflected in the removal rate of the CMP process through the chemical/physical synergism involved in the shear-enhanced desorption of the altered SiLK surface layer.

The five step mechanism for CMP indicates that the velocity and pressure have a contributing effect to the shear-enhanced desorption step (iv) as well as the mass-transport steps (i and v). There exists a functional dependence of the reaction rate constant for desorption on the adjustable parameters of the CMP system, such that:

$$k_{des} = f(P, U, c_a, d_p, T) \tag{6.30}$$

Equation 6.30 states that $k_{des}$ is some function of the CMP pressure ($P$) and velocity ($U$), the abrasive particle size ($c_a$), the particle diameter ($d_p$), and the system temperature ($T$). Additional experiments were performed with constant $c_a$, $d_p$, and $T$, while varying pressure and velocity to examine their effects.

To further investigate the dependence of $K_1$, and thus $k_{des}$ on the physical parameters of the CMP system, the SiLK removal rate was measured for 12 combinations of velocity and pressure, while keeping the slurry concentration constant at $A_{active}|_{inlet} = 0.012$ M. The data are tabulated in Appendix C. The model was then used to calculate the measured removal rate by changing the input pressure and velocity, and adjusting $K_1$ to fit the experimental removal rate data. The data are plotted as $K_1$ vs. $sqrt(UP)$, since $sqrt(UP)$ is proportional to the shear stress generated during a lubrication flow[6.2]. The values of $K_1$ required to match the experimental data are shown in Figure 6.23.

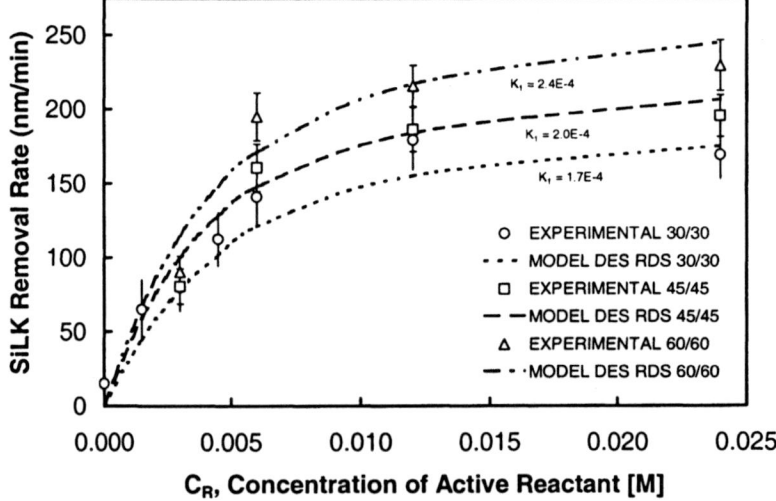

*Figure 6.22.* Model predictions showing variation in SiLK CMP removal rate with CMP velocity and variable $K_1$ ($\alpha\, k_{des}$), for the case of shear-enhanced desorption RDS.

Very good model fits for the removal rate at varied P and U have been calculated by varying $K_I$ in Equations 6.26 and 6.27. Moreover, the power-law data fit obtained from Figure 6.23 suggests a correlation dependence of $K_I$ (and thus $k_{des}$) on the pressure and relative velocity at the wafer surface. Changes in velocity are represented in both a chemical (mass transport), and physical (shear-stress aided removal) manner. Therefore, Equation 6.30 is modified to conform to the observed relationship between $k_{des}$, P, and U, which can be expressed by one of two correlation equations:

power-law $\quad k_{des} = 1.2 \times 10^{-5} [(UP)^{0.5}]^{2.3} f(c_a, d_p, T)$ (6.31)

exponential $\quad k_{des} = 0.23 exp[0.02(UP)^{0.5}] f(c_a, d_p, T)$ (6.32)

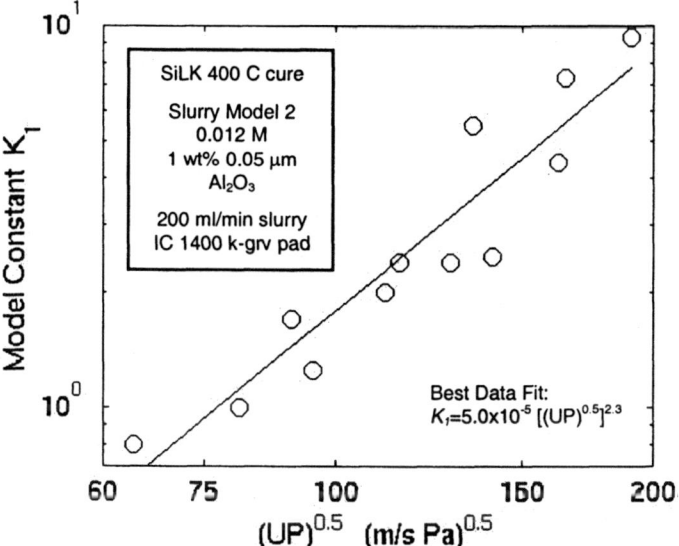

*Figure 6.23.* Correlation of $K_I$ and $(UP)^{0.5}$; representing the functional dependence of the desorption constant, kdes, on the fluid shear stress at the wafer surface during SiLK CMP.

Figure 6.23 and Equation 6.30 show that there is a dependence of $k_{des}$ on the physical parameters of the CMP system that is more complex than a direct proportionality to the pressure, velocity, or shear stress. There also exists a dependence of $k_{des}$ on the pad and abrasive particle contact mechanics, which determines the value of the rate constant for shear-enhanced desorption. Correlating $k_{des}$ to CMP process variables illustrates the main virtue of the multistep reaction kinetics model – that the adjustable parameter (in this case $K_{des}$, which is proportional to $k_{des}$), may be examined for dependence on any CMP process variable. By re-calculating the model results as a process variable is changed and examining the functionality of the adjustable parameter, one can determine the relationship between the process variable and the rate constant of a particular reaction step, providing additional insight into the mechanisms for material removal.

Pressure ($P$), velocity ($V$), particle diameter ($d_p$), and slurry reactant concentration ($A_{active|inlet}$) are examples of process variables that, when changed, result in measurable differences in CMP removal rate. However, changing the CMP parameters sometimes causes deviation from the synergistic chemical/mechanical removal model, resulting in the domination of either chemistry or physical removal, and either extremely low removal rate or high removal rate with extensive film scratching. Some process conditions that result in experimentally measured SiLK removal that deviate from the synergistic altered-layer model are listed in Appendix C. Most notably, scratch deviations occur when $d_p > 50$ nm, $P \geq 5.5$ psi, and $A_{active|inlet} = 0$ mol/m$^3$.

After successfully matching model predictions with the experimental data for varied velocity and pressure, we proceeded to test the ability of the model to achieve a mass-transport-limited case. Figure 6.21 indicates that mass transport effects alone are not sufficient to account for the increases observed in removal rate with velocity. However, the model framework may be used to understand more about the CMP system by forcing the system to exhibit mass-transport dependence and varying the CMP parameters to understand their impact.

When surface reaction controls the rate of removal, then Equation 6.16 is used as the wafer surface boundary condition, with leading constant $K_1 \propto k_{rxn}$. When made sufficiently large, $k_{rxn}$ causes the reaction in Equation 6.6 to shift to the right. This shift also causes rapid consumption of the reactant at the wafer surface and a large gradient between the reactant concentration at the wafer surface and the bulk slurry. Since the removal rate is now dependent on the arrival of chemical to the surface (due to a very fast surface chemical reaction), mass transport of the reactant to the surface is the overall rate determining step for the CMP process. For this specific case, increasing the velocity (and thus the mass transport of species to and away from the wafer surface) results in large increase in the removal rate, because the mass transport resistance is decreased. Results are shown in Figure 6.24.

Figure 6.24 contains model results that were generated for six different velocity combinations, using a model constant, $K_1$ ($75 \times 10^{-4}$ mol/m$^2$ s), two orders of magnitude greater than the "best fit" value ($1.4 \times 10^{-4}$ mol/m$^2$ s) used to calculate the model fit for reaction RDS in Figure 6.20. The larger value of $K_1$ causes rapid reactant consumption at the wafer surface, and results in values for the removal rate which are much higher than the experimental values. The comparison of these results is most useful for model testing rather than reproduction of experimental data. Figure 6.24 shows that in the case of rapid surface processes, mass transport is indeed rate limiting, and

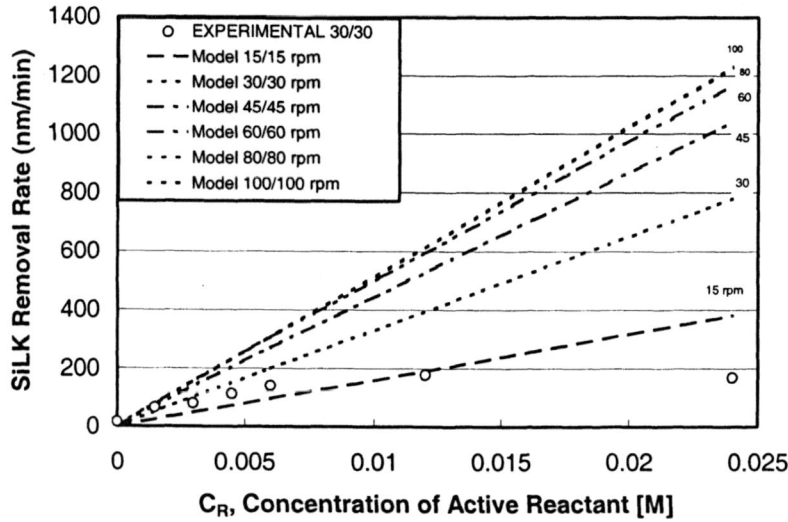

*Figure 6.24.* Variation in removal rate for mass-transport limited case, when surface processes are fast and reactant is consumed rapidly. Pressure = 2.5 psi.

removal rate can be increased greatly by varying the CMP velocity. This is illustrated in Table 6.4. The removal rate is proportional to velocity at low velocity. The increase in removal rate with velocity approaches an asymptote as the platen and carrier velocities approach 100 rpm. The results suggest that the model can capture the effects of mass-transport resistance, but that these effects are not rate controlling in the case of SiLK CMP at the slurry concentrations and tool operating parameters (velocity, pressure) that have been examined experimentally.

*Table 6.4.* Variation of model removal rate in mass-transport limited regime, for two slurry reactant concentrations (0.006 M and 0.024 M). Pressure = 2.5 psi.

| $C_{Active|inlet}$ = 0.006 M | | | $C_{Active|inlet}$ = 0.024 M | | |
|---|---|---|---|---|---|
| Velocity (RPM) | $RR_{calc}$ (nm/min) | % increase | Velocity (RPM) | $RR_{calc}$ (nm/min) | % increase |
| 15 | 96 | -- | 15 | 380 | -- |
| 30 | 200 | 108 % | 30 | 780 | 104 % |
| 45 | 270 | 36 % | 45 | 1040 | 33 % |
| 60 | 310 | 13 % | 60 | 1170 | 12 % |
| 80 | 310 | ~ 0 % | 80 | 1230 | 5 % |
| 100 | 310 | ~ 0 % | 100 | 1230 | ~ 0 % |

The model results provide valuable information about the mechanism for SiLK CMP. The results show that the surface processes (reaction, desorption) are rate-limiting, rather than mass transport under the experimental conditions used. They also show the interaction or synergism between the slurry chemistry (surface reaction) and physical forces (shear-enhanced removal of the reacted layer), through their impact on the rate constants $k_{rxn}$ and $k_{des}$.

## 6.5 TWO STEP REMOVAL MODEL FOR SILK CMP USING HETEROGENEOUS CATALYSIS

In light of the experimental data and the 5-step Langmuir-Hinshelwood model, a simplified two-step mechanism can be used to predict the interaction between the SiLK removal rate and the slurry reactant concentration [6.20]. Figures 6.2 and 6.5 show that two regimes exist for SiLK removal rate when the slurry concentration is varied; a low-concentration linear regime where slurry chemistry is more important, and a high concentration asymptotic regime where slurry chemistry becomes less of a factor. A two-step model captures the fundamental processes in these two regimes -- (1) a surface reaction step and (2) a shear-enhanced removal step -- in order to predict CMP removal rate as a function of reactant concentration.

### 6.5.1 Two Step Surface Mechanism

A two step reaction-based model can be used to correlate the SiLK removal rate as a function of the concentration of the active chemical species present in the slurry (from previous experiments, this can be either QCTT1010 chemical or KH phthalate). The model incorporates the steps shown to be most important in the more general five-step treatment of the CMP process. These steps are surface reaction and mechanically-assisted adsorption.

In step 1 of the mechanism, the material to be polished $\underline{M}$, reacts with the active chemical species in the slurry $R$, to form a reacted layer on the wafer surface $\underline{L}$. For example, in the case of SiLK CMP, $\underline{M}$ is a SiLK polymer surface site, $R$ is KH phthalate, and $\underline{L}$ is a reacted SiLK surface site.

$$\underline{M} + R \xrightarrow{k_1} \underline{L} \qquad (6.33)$$

In the above equation, $k_1$ is the surface reaction rate constant, and has units of m/[conc]s. The under-bars denote species present on the wafer surface. Assuming that an irreversible surface-conversion reaction, the rate of formation of $\underline{L}$ is given by:

$$r_1 = k_1 \theta_M [R] \tag{6.34}$$

where $\theta_M$ is the fraction of SiLK surface sites that are available on the wafer surface and $[R]$ is the concentration of the reactant in the slurry. In this mechanism for the CMP process, mass transport and adsorption resistance of $R$ to the wafer surface are assumed to be fast and offer negligible resistance to removal. This is a reasonable simplification, since adsorption is generally a very fast phenomenon.

In step 2 of the mechanism, the reacted layer $\underline{L}$ is removed from the wafer surface by abrasion and appears in the bulk slurry (denoted by $L$):

$$\underline{L} \xrightarrow{k_2} L \tag{6.35}$$

where $k_2$ is the altered surface desorption/removal rate constant, and has units of m/s. It is reasonable to assume that abraded material will not re-deposit on the surface and again become a chemically-bound section of the SiLK structure. In addition, the mass transport of the abraded material from the wafer surface to the slurry bulk is assumed to be rapid. We note that $k_2$ is likely to be a function of the stress at the wafer surface as well as abrasive size and concentration.

The rate of formation of $L$, the abraded SiLK product, is given by:

$$r_2 = k_2 \theta_{\underline{L}} \tag{6.36}$$

where $\theta_{\underline{L}}$ is the fraction of the wafer surface where $\underline{L}$ is present and $k_2 = k_2(\tau_w, C_a, d_p)$. $\tau_w$ is the shear stress at the wafer surface, $C_a$ is the concentration of abrasive particles in the slurry, and $d_p$ is the diameter of the abrasive particles.

If the model is in steady state, then $r_1 = r_2$, and the rate of $\underline{L}$ going into the bulk ($r_2$) is equal to the material removal rate ($RR$). We can solve for $r_2$ in terms of $[R]$ using a balance of sites on the wafer:

$$\theta_{\underline{M}} + \theta_{\underline{L}} = 1 \tag{6.37}$$

Assuming a steady state for the intermediate species $\underline{L}$ we obtain

$$RR = r_1 = r_2 \tag{6.38}$$

Equations 6.34 and 6.36 - 6.39, give an expression for the SiLK removal rate as a function of the reactant concentration and the rate constants for surface reaction and altered-layer desorption:

$$RR = \frac{k_1 [R]}{1 + \frac{k_1}{k_2}[R]} \tag{6.39}$$

## 6.5.2 Results

Equation 6.39 can be used to correlate the experimental data for SiLK removal rate versus QCTT1010 or KH phthalate concentration (with different units for the constant $k_1$, since the slurry concentrations have different units). The data are correlated by fitting Equation 6.39 to the experimental data in Figures 6.2 and 6.5. The two-step removal mechanism model fits are generated by solving for the best fit combination of $k_1$ and $k_2$, which represent the forward rate constants for (i) surface reaction and (ii) mechanical removal of the altered surface layer.

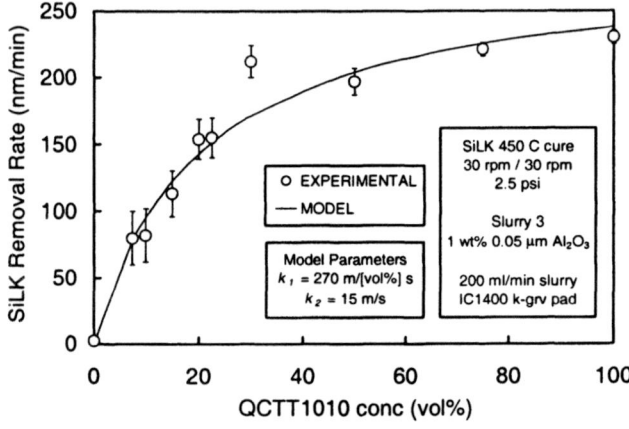

*Figure 6.25.* Two step surface reaction model fit to experimental data for SiLK removal rate versus QCTT1010 chemical concentration (from ref[6.20]).

Figures 6.25 and 6.26 show that the two-step removal rate model based on heterogeneous reaction kinetics can correlate SiLK CMP removal over a varied concentration of slurry reactant [6.20]. These results support the hypothesis that two regimes control SiLK/carboxylic acid CMP. At low slurry concentrations the removal rate increases linearly with concentration since the CMP process is surface reaction rate limited as $RR \sim k_1[R]$. At high slurry concentrations the removal rate plateaus to a maximum value as the surface reaction becomes saturated, and the CMP process becomes shear/abrasion rate limited as $RR$ approaches $k_2$.

The model correlations can be used to select the proper slurry chemical concentration to achieve maximum removal rate without needlessly over-concentrating the slurry. Although the model may be used to predict removal rate results at high and low chemical concentrations, the absolute values of the model parameters $k_1$ and $k_2$ may not be compared directly

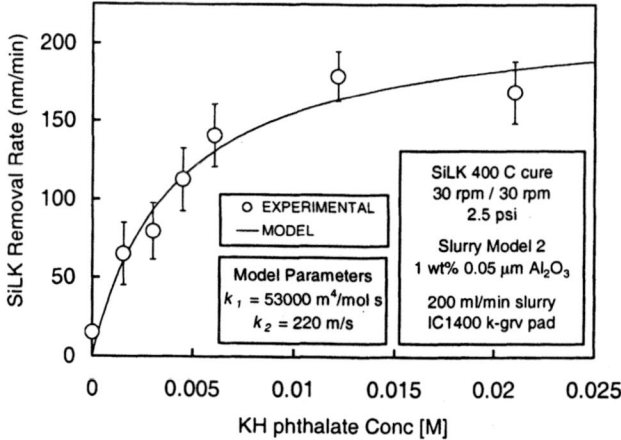

*Figure 6.26.* Two step surface reaction model fit to experimental data for SiLK removal rate versus KH phthalate chemical concentration (from ref[6.19]).

unless they have the same units. However, since $k_1$ is the "chemical" rate constant and $k_2$ is the "mechanical" rate constant, a higher $k_1$ (with the same units for comparison) denotes more rapid chemical alteration of the surface, and a higher $k_2$ (with the same units for comparison) denotes a slurry which more aggressively removes the polymer surface layer.

## 6.6 EXTENDIBILITY OF MODEL TO DESCRIBE THE CMP OF OTHER MATERIALS

Although the multi-step surface kinetics model has been developed and tested for SiLK polymer removal, the generic Langmuir-Hinshelwood form can be applied to various materials and CMP slurries. For application to other materials, the following information is desired:
1) basic knowledge of film structure and active slurry chemical reactant
2) density, molecular weight, diffusivity, and stoichiometry information for the material to be polished (for *RR* Equation 6.28)
3) estimates for equilibrium constants for adsorption, reaction, and desorption for the slurry chemistry and the material to be polished (for B.C. Equations 6.26 and 6.27)

With this information, the model can be used to calculate removal rates, fit predictions to experimental data, and determine information regarding the governing mechanisms for the removal of materials other than SiLK.

### 6.6.1 Copper

With the recent shift from aluminum to copper conductor, methods for modeling the CMP of copper have become increasingly important. A mechanistic model such as the one discussed in sections 6.4 and 6.5 can be used to forward the understanding of copper CMP mechanisms. CMP experiments have been performed by Lee et al.[6.21] with a simplified copper slurry (Figure 6.27). The use of potassium dichromate ($K_2Cr_2O_7$) oxidizer results in a trend for removal rate versus slurry composition that is very similar to the trend observed with SiLK and KH phthalate. The two-step mechanistic surface-reaction model captures the phenomena that affect copper CMP. In Chapter 7, a modified L-H approach similar to that used in Section 6.4 is extended to model copper CMP, based on the work of Thakurta et al.[6.22]. The approach focuses on the three-dimensional aspects of the CMP process and slurry mass-transport considerations.

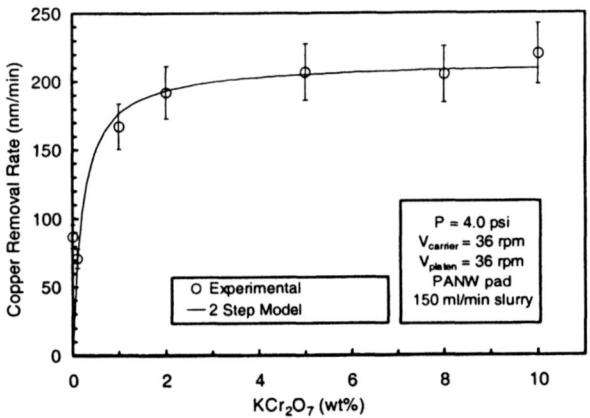

*Figure 6.27.* Data for the CMP removal of copper using a simple DI water and potassium dichromate ($KCr_2O_7$) slurry. Equation 6.39 correlates data from ref[6.20].

### 6.6.2 Dielectrics

The multi-step surface reaction kinetics model also shows promise for modeling the CMP of dielectric materials. Figure 6.28 shows experimental data for silicon dioxide removal versus dilution of a commercial potassium-hydroxide (KOH) slurry chemistry. This data must be interpreted carefully, since the dilution of the slurry results in reduced concentration of slurry chemistry and slurry abrasive particle concentration. But the data in the figure shows a trend similar to that observed with SiLK and copper CMP. There exists a concentration of the active component in the slurry above which the removal rate of material does not change dramatically.

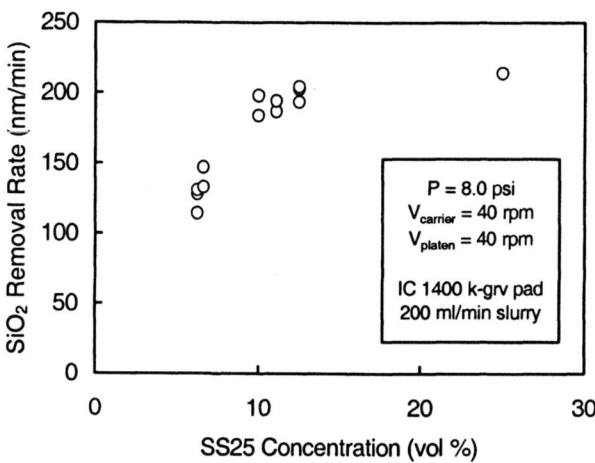

*Figure 6.28.* Data for the CMP removal of TEOS $SiO_2$ using a diluted commercial KOH chemistry.

This "plateau" or asymptote reflects the transition from reaction-rate dependence to desorption-rate dependence. Similar data are shown in Figure 6.29 for the CMP removal of FLARE low-κ material versus slurry concentration of ferric nitrate ($FeNO_3$) oxidizer.

The types of data shown in the figures can be used to better understand the mechanisms for dielectrics of all kinds. Selection of the most effective slurry chemistry (chemical concentrations and optimal abrasive content) to achieve desired CMP properties (removal rate, selectivity, and film surface finish) can be obtained using the techniques presented here.

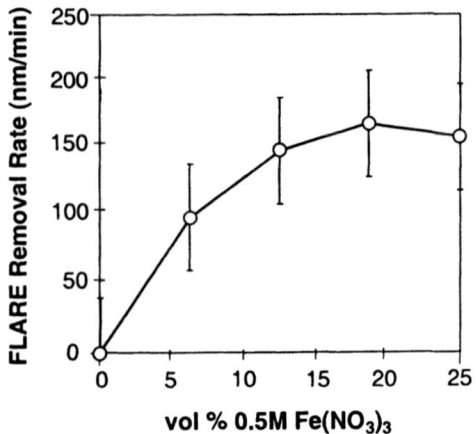

*Figure 6.29.* Data for the CMP removal of FLARE low-κ dielectric using ferric nitrate ($Fe(NO_3)_3$) slurry chemistry (from ref[6.4])

## 6.7 REFERENCES

[6.1]  S. Sundararajan, D.G. Thakurta, D.W. Schwendeman, S.P. Murarka, W. N. Gill, *J. Electrochem. Soc.*, **146(2)**, (1999).
[6.2]  D.G. Thakurta, C. L. Borst, D. W. Schwendeman, R.J. Gutmann, W. N. Gill, *Thin Solid Films*, **366**, 181 (2000).
[6.3]  D.G. Thakurta, C.L. Borst, D.W. Schwendeman, R.J. Gutmann, W.N. Gill, submitted to *J. Electrochem. Soc.* (2000).
[6.4]  D. Towery and M. Fury, *J. Elect. Mat.*, **27(10)**, 1088 (1998).
[6.5]  J.M. Smith, Chemical Engineering Kinetics, Second Edition, New York: McGraw-Hill Book Company (1970).
[6.6]  V. Nguyen, H. VanKranenburg, P. Woerlee, *Microelectron. Eng.*, **50**, 403 (2000).
[6.7]  B. Jirgensons, M.E. Straumanis, A Short Textbook of Colloid Chemistry, New York: MacMillan (1962).
[6.8]  J.G. Dos Ramos, C.A. Silebi, *Polym. Int.*, 30 (1993).
[6.9]  C.L. Borst, D.G. Thakurta, R.J. Gutmann, W.N. Gill, *J. Electrochem. Soc.*, **146(11)**, 4309 (1999).
[6.10] J.M. Neirynck, S. P. Murarka, R. J. Gutmann, in: T.-M. Lu, S.P. Murarka, T.-S. Kuan, and C.H. Ting, *Low-Dielectric Constant Materials - Synthesis and Applications in Microelectronics*, San Francisco, USA, April 17-19, 1995, Materials Research Society Symposium Proceedings, **381**, 229 (1995).
[6.11] J.M. Neirynck, G.-R. Yang, S.P. Murarka, R.J. Gutmann, *Thin Solid Films,* **290-291**, 447 (1996).
[6.12] D. Permana, S. P. Murarka, M. G. Lee, S. I. Beilin in: R. Havemann, J. Schmitz, H. Komiyama, K. Tsubouchi, *Advanced Metallization and Interconnect Systems for ULSI Applications in 1996*, Boston, USA, October 1-3, 1996, Proceedings of Advanced Metallization and Interconnect Systems for ULSI Applications in 1996, 539 (1997).

[6.13] G.-R. Yang, Y.-P. Zhao, J. M. Neirynck, S. P. Murarka, R. J. Gutmann, *J. Electrochem. Soc.*, **144(9)**, 3249 (1997).
[6.14] L. Cook, *J. Non-Cryst. Solids*, **120**, 152 (1990).
[6.15] C. L. Borst, D. G. Thakurta, W. N. Gill, R. J. Gutmann, *J. Electrochem. Soc.*, **149(2)**, G118 (2002).
[6.16] S. W. Waite, J. F. Holzwarth, J. M. Harris, *Anal. Chem.*, **67(8)**, 1390 (1995).
[6.17] F. Y. Ren, J. M. Harris, *Anal. Chem.*, **68(9)**, 1651 (1996).
[6.18] J. H. Penn, Z. Lin, *J. Org. Chem.*, **55(5)**, 1554 (1990).
[6.19] R. L. Hansen, J. M. Harris, *Anal. Chem.*, **70(20)**, 4247 (1998).
[6.20] R. J. Gutmann, C. L. Borst, B.-C. Lee, D. Thakurta, D. Duquette, W. N. Gill, *17th VMIC Conference*, Santa Clara, CA, June 27-29, 123 (2000).
[6.21] B. C. Lee, B. Wang, D. J. Duquette, R. J. Gutmann, *4th Intl. Conf. on CMP Planar. (CMP-MIC) Conference*, March 2-3, Santa Clara, CA (2000).
[6.22] D. G. Thakurta, D.W. Schwendeman, R.J. Gutmann, S. Shankar, L. Jiang and W.N. Gill, Thin Solid Films, accepted for publication, 2002.

# Chapter 7

# COPPER CMP MODEL BASED UPON FLUID MECHANICS AND SURFACE KINETICS

The trend to replace aluminum with copper in IC interconnects requires that CMP be used to planarize the copper which has been deposited into RIE-patterned structures. Therefore, in this chapter, we discuss and develop a copper CMP model which predicts wafer-scale removal rate based on mass-transport theory and surface kinetic steps. The model seems to capture the essential features of the planarization of copper, and is very similar to the surface reaction kinetics model developed for low-κ CMP in Chapter 6. The model computes the concentration of the chemical reactant in the slurry using a convective diffusive mass transport equation. Surface kinetic equations are used to model the chemical reaction and mechanical abrasion processes at the wafer surface during CMP. The model approach for the CMP of copper is based upon a recent paper [7.1] and a PhD thesis [7.2]. The approach used here incorporates the fundamentals of pad/abrasive contact into the lubrication and mass transport process. It extends upon the low-κ CMP model developed previously, and may find applicability to dielectrics and other metals with well-known chemical reaction kinetics.

The model is built around the geometry and flow that are characteristic of a conventional rotary CMP system as depicted in Figure 7.1. The slurry, containing chemical reagents and abrasive particles, is drawn beneath the wafer by the rotating pad. The slurry forms a lubricating film between the wafer and pad, with an average film thickness, $h$, between 20 and 60 μm [7.2]. Figure 7.1(b) shows a side profile of the pad surface which has asperities of randomly varying heights [7.3-7.5]. The surface roughness of the pad, defined as the standard deviation of the height of the asperities, ranges from 10 to 30 μm depending on the pad properties [7.6-7.7]. With $h$ of the same order as the mean height of the asperities, a significant fraction of the asperities are in contact with the wafer surface. The applied load on the wafer is carried by the hydrodynamic pressure developed in the slurry film and by the asperities in contact with the wafer.

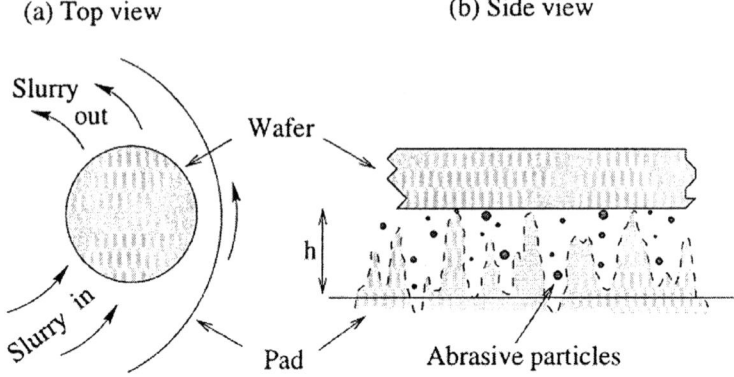

*Figure 7.1:* Schematic top and side view of wafer and pad.

For purposes of fluid flow modeling, the slurry is assumed to flow between the finger-like protrusions (asperities) on the pad surface. Instead of considering the slurry flow through the asperities [7.8] using statistical averaging techniques [7.9], the 3-D wafer-scale slurry flow model developed by Thakurta et al. [7.10] which treats the pad as a smooth surface is employed. In particular, the effect of contact of the asperities with the wafer is brought into the mass-transport model through a boundary condition describing the surface kinetics at the wafer surface. The flux of the reactant at the wafer surface is computed from its concentration distribution, which is related to the local removal rate through a modified Langmuir-Hinshelwood model developed for copper CMP [7.2], in parallel with the CMP model presented in Chapter 6 for low-κ dielectrics.

Various investigators have developed models for copper chemical-mechanical planarization. Sundararajan et al. [7.11] analyzed the mass transport in copper CMP using a 2-D lubrication model to analyze the mass-transport in the slurry film, which was coupled to a chemical reaction at the wafer surface. The model agrees reasonably well with experiments, predicting the effects of slurry chemistry, relative speed between the wafer and pad, and applied pressure on the wafer. The present description includes mass transport and flow models in 3-D that enable prediction of the wafer-scale local material removal rate using a more complete surface reaction kinetics scheme.

Subramanian et al. [7.12] have investigated transport phenomena issues in CMP using a different approach. The pores of the pad are modeled as rectangular cells which carry the slurry underneath the wafer. A mass-transport equation is used to describe the reactant concentration distribution in the cell, and the flux at the wafer surface is used to obtain the material removal rate. In contrast, the transport model presented here considers the pad fibers as protrusions from the pad surface which do not restrict the slurry flow within a cell. Also an additional kinetic rate parameter is introduced which accounts for mechanical abrasion at the wafer surface.

Since the flow field has an important effect on mass transfer, this chapter begins with a summary of the flow model of Thakurta et al. [7.10]. In subsequent sections, equations that describe (1) mass-transport in the gap between the wafer and (2) the pad and the boundary condition given by our surface kinetics model (which includes both chemical and mechanical effects) are developed. The flux of copper at the wafer surface is determined, and this flux integrated to obtain the removal rate of copper. An effectiveness factor is defined which measures the mass-transport resistance to the overall removal process, and determines which aspect of CMP controls the removal process: (1) mass transfer in the slurry, or (2) surface chemical and mechanical removal. The kinetic rate parameters used in the model are then discussed, followed by a discussion of the numerical solution procedure. Results of the model for a representative flow configuration then are presented, followed by a summary of experimental results used to test the model and determine the kinetic parameters.

## 7.1 Flow Model

The purpose of the flow model is to compute the slurry film thickness and the velocity distribution of the slurry in the gap between the wafer and pad. The magnitude and distribution of the gap between the pad and wafer are determined by CMP operating parameters such as wafer and pad velocity and applied pressure. The 3-D wafer-scale flow model developed by Thakurta et al. [7.10] is briefly described in this section. The model uses CMP pressure and velocity parameters as input to calculate the fluid gap height, $h(x,y)$ and the fluid velocity distribution $v(x,y)$ throughout the gap. These computed quantities are required as inputs to the mass-transport model that calculates the copper surface removal.

The geometry of the wafer-pad system is shown in Figure 7.2. The wafer radius is $R_1$ and the distance between the centers of the wafer and pad is $R_2$. The wafer and pad rotate about their respective centers with angular speeds $\omega_1$ and $\omega_2$, respectively. The coordinate system is fixed in space with the origin at the pad surface directly below the center of the wafer. The wafer height given by $z = h(x,y)$ is measured relative to the pad surface (the $x$-$y$ plane) and includes a global convex curvature and angles of attack. The curvature is represented in terms of the protrusion at the center of the wafer, or the wafer dome height [7.10]. The wafer surface is taken to be

$$h(x, y) = h_0 + S_x \left[\frac{x}{R_1}\right] + S_y \left[\frac{y}{R_1}\right] + \delta_0 \left[\frac{x^2 + y^2}{R_1^2}\right] \quad (7.1)$$

where $h_0$ is the wafer height at the origin, and $S_x$ and $S_y$ are slopes associated with the angles of attack in the $x$- and $y$-directions, respectively. The parameters $h_0$, $S_x$ and $S_y$ are adjusted to balance the applied pressure and to satisfy zero moments on the wafer about the center [7.3, 7.10]. As mentioned, these parameters depend on pressure and velocity.

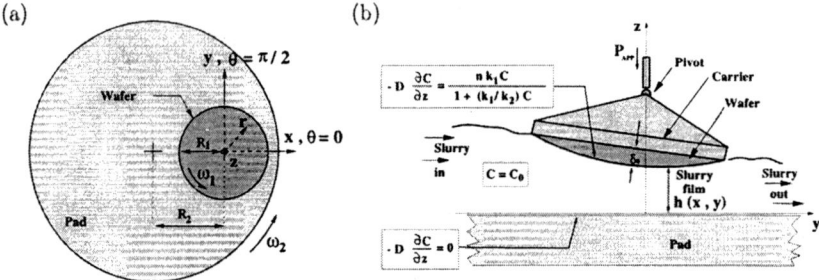

*Figure 7.2:* (a) Schematic top view of CMP tool showing pad and wafer positions. (b) Schematic side view of 3-D model. Boundary conditions (for the mass-transport model) at inlet, pad and wafer surfaces are shown within boxes.

The flow model considers the lubrication approximation of the Navier-Stokes equations, which is solved numerically to obtain the slurry pressure in the gap and the adjustable parameters in the expression for $h$. The velocity distribution in the fluid gap is computed based on the slurry pressure calculation. Under typical CMP conditions, the computed slurry film thickness ranges from 20 to 60 μm, and the slurry flow speeds range from 0.3 to 1.4 ms$^{-1}$, as presented in detail elsewhere [7.2, 7.10].

## 7.2 Copper Removal Model

The copper removal model developed by Thakurta et al. [7.1] is presented here in three steps: (1) mass transport of oxidizer to the wafer surface, (2) chemical reaction of the oxidizer with copper to form a reacted layer at the wafer surface, and (3) removal of both (a) the reacted layer and (b) copper by mechanical abrasion. These steps are depicted in Figure 7.3 and are discussed below.

# COPPER CMP MODEL BASED UPON FLUID MECHANICS AND SURFACE KINETICS

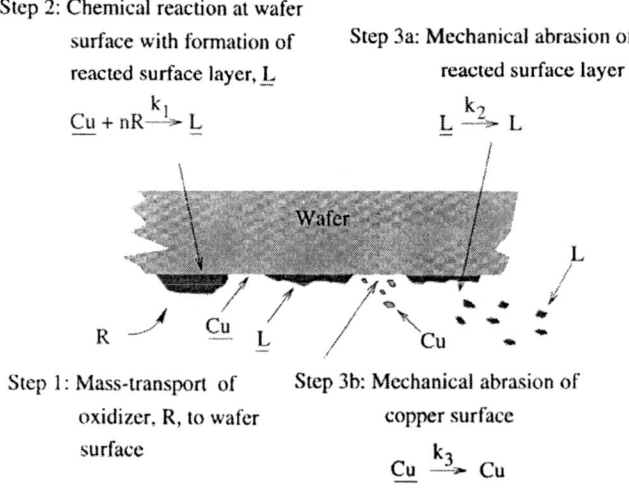

*Figure 7.3:* Steps in the removal model for copper.

## 7.2.1 Mass Transport of Oxidizer to The Wafer Surface

The fresh slurry consisting of an oxidizer, $R$, enters the thin slurry-film region between the wafer and pad as shown in Figure 7.1(a). The copper at the wafer surface, $\underline{Cu}$, reacts with n moles of $R$ forming a reacted layer, $L$, which then is removed from the surface by mechanical abrasion. The overall reaction of copper with the oxidizer may be written as follows:

$$\underline{Cu} + nR \to \underline{L} \qquad (7.2)$$

where the symbols with under-bars denote species present on the wafer surface [7.14]. The slurry leaves the region depleted in $R$ and enriched in $L$. The rate of material removal depends upon the availability of $R$ in the fluid phase at the wafer surface. Both convection and diffusion transport $R$ to the wafer surface. Mass transport of $R$ to the surface is Step 1 in the material removal process (see Figure 7.3). The mass-transport model is solved to obtain the concentration of $R$ between the wafer and pad. The concentration distribution is used to compute the rate of consumption of $R$ at the wafer surface and, subsequently, the local copper removal rate.

Assuming a dilute system with constant liquid density and diffusivity, we can write the following governing equation for the concentration of oxidizer $R$, $C(r,\theta,z)$, in cylindrical coordinates:

$$v_r \frac{\partial C}{\partial r} + v_\theta \frac{1}{r}\frac{\partial C}{\partial \theta} + v_z \frac{\partial C}{\partial z} = D \frac{\partial^2 C}{\partial z^2} \qquad (7.3)$$

where $D$ is the diffusivity of $R$ in the slurry and $v_r$, $v_\theta$, $v_z$ are functions of $(r,\theta,z)$ given by the slurry flow model. In Equation 7.3, the contribution to diffusion in $r$- and $\theta$-directions is neglected because $h \ll R_1$. The equation for $C$ is solved in the region between the wafer and pad given by $r \in [0,R_1]$, $\theta \in 0[0,2\pi]$, and $z \in [0, h]$, with appropriate boundary conditions at inlet, pad and wafer surfaces. The inlet is defined as the regions at $r = R_1$ where the slurry, assumed 'fresh' in R, enters the space between the wafer and pad. Thus, the inlet boundary condition for $C$ is given by

$$C(R_1,\theta,z) = C_0 \quad \text{when} \quad v_r(R_1,\theta,z) < 0 \qquad (7.4)$$

where $C_0$ is the molar concentration of $R$ in fresh or as prepared slurry (assumed to be known), and the velocity criterion in Equation 7.4 identifies the inlet regions. Since the pad is considered to be inert and impermeable, the diffusive flux is set to zero at the pad surface

$$-D\frac{\partial C}{\partial z}(r,\theta,0) = 0. \qquad (7.5)$$

### 7.2.2 Surface Kinetic Steps

The mass transport of $R$ is followed by Steps 2 and 3 both occurring at the wafer surface (see Figure 7.3). In Step 2, the copper film, $\underline{Cu}$, reacts with n moles of $R$ in the fluid phase to form a reacted layer, $\underline{L}$, on the wafer surface:

$$\underline{Cu} + nR \xrightarrow{k_1} \underline{L} \quad \text{Step 2,} \qquad (7.6)$$

assuming that the adsorption of R on the copper surface is sufficiently fast and that the backward reaction rate of Equation 7.6 is negligible. The reacted surface layer, $\underline{L}$, is an oxidized copper compound depending upon $R$. In this model, part of the wafer surface is covered by the reacted layer, $\underline{L}$, while the rest is unreacted copper, $\underline{Cu}$. In Step 3, topographically higher regions of the reacted surface layer are mechanically removed by a combined action of the pad and abrasive particles, and the abraded material, $L$, goes into the slurry (Step 3a). Direct removal of the copper surface by mechanical abrasion occurs simultaneously (Step 3b). The model assumes that the abraded material does not re-deposit onto the wafer surface. The mechanical abrasion steps are given as follows:

$$\underline{L} \xrightarrow{k_2} L \quad \text{Step 3a.} \qquad (7.7)$$

$$\underline{Cu} \xrightarrow{k_3} Cu \quad \text{Step 3b.} \qquad (7.8)$$

Equations 7.6 and 7.8 show that copper is removed by two mechanisms. In Equation 7.6 copper is depleted by chemical transformation to $\underline{L}$, while in Equation 7.8 copper is simultaneously removed by mechanical abrasion. Overall, the molar rate of removal of copper is given by the rate of change of

# COPPER CMP MODEL BASED UPON FLUID MECHANICS AND SURFACE KINETICS

its mass on the surface divided by the molecular weight of copper, $\frac{1}{MW}\frac{dm}{dt}$. The mass of the copper film on the wafer is the product of the density of copper and the volume of the film, $\rho V$. The volume V is the product of the average film thickness and the film area. Thus $\frac{dm}{dt}=A\rho\frac{df}{dt}$ where $\frac{df}{dt} = RR(r,\theta)$, the removal rate.

The chemical reaction in Step 2 and the mechanical abrasive actions in Steps 3a and 3b are described by the rate parameters $k_1$, $k_2$ and $k_3$, respectively. The rate parameter $k_1$ is determined by the chemistry of reactant R with copper and is assumed to be constant independent of the CMP operating conditions. The rate parameters $k_2$ and $k_3$, on the other hand, are dependent on a number of CMP parameters, in particular the pad and wafer speed and applied pressure.

A first order chemical reaction with respect to R with a kinetic rate parameter $k_1$ is assumed in Equation 7.6. The local rate of formation of $\underline{L}$ per unit surface area by Step 2, $r_1$, is obtained by using the principles of kinetics of fluid-solid catalytic reactions [7.13]. The rate is proportional to the product of the concentration of R in the fluid phase at the wafer surface and the probability that the surface site is $\underline{Cu}$

$$r_1 = k_1 \theta_{\underline{Cu}} C(r,\theta,h), \qquad (7.9)$$

where $\theta_{\underline{Cu}}$ is the probability that the surface site is $\underline{Cu}$ and $C(r, \theta, h)$ is the local concentration of R in the slurry at the wafer surface. Similarly, the rates (per unit surface area) of $\underline{L}$ and $\underline{Cu}$ going into the slurry (by Steps 3a and 3b, respectively) are given by

$$r_2 = k_2 \theta_{\underline{L}}, \text{ and} \qquad (7.10)$$

$$r_3 = k_3 \theta_{\underline{Cu}}, \qquad (7.11)$$

respectively; $\theta_{\underline{L}} = 1-\theta_{\underline{Cu}}$ is the probability that the surface site is $\underline{L}$. Note that $r_1$ is a chemical reaction rate which is proportional to the concentration of the active component (oxidizer) at the wafer-slurry interface. In contrast, $r_2$ and $r_3$ refer to the rates of a mechanical removal process, and as such are not proportional to the oxidizer. They are functions of the mechanical parameters of the system which determine $k_2$ and $k_3$.

To deduce the fractions of the surface covered by copper and $\underline{L}$, it is assumed that the rate of formation of $\underline{L}$ is equal to its rate of depletion, which gives

$$r_1 = r_2. \qquad (7.12)$$

Substituting Equations 7.9 and 7.10 into Equation 7.12 and solving for $\theta_{\underline{Cu}}$ gives

$$\theta_{\underline{Cu}} = \cfrac{1}{1+\cfrac{k_1}{k_2}C(r,\theta,h)}. \qquad (7.13)$$

Equation 7.13 indicates that the fraction of the surface covered by copper is inversely related to the oxidizer concentration. The above expression of $\theta_{\underline{Cu}}$ may be substituted back in Equation 7.9 to obtain $r_1$ and $r_2$ (which are the rates of formation and depletion of the altered layer $\underline{L}$) in terms of the kinetic parameters and the wafer surface concentration of $R$:

$$r_1 = r_2 = \cfrac{k_1 C(r,\theta,h)}{1+\cfrac{k_1}{k_2}C(r,\theta,h)}. \qquad (7.14)$$

From stoichiometry of the reaction in Equation 7.6, the flux of the oxidizer (or active slurry component) $R$ normal to the wafer surface is determined by the transformation of copper into $\underline{L}$, giving

$$-D\frac{\partial C}{\partial z}(r,\theta,h) = nr_1 = \cfrac{nk_1 C(r,\theta,h)}{1+\cfrac{k_1}{k_2}C(r,\theta,h)}. \qquad (7.15)$$

Equation 7.15 indicates how the flux of R and the rate of formation of $\underline{L}$ are related to one another. The structure is very similar to the two-step model formulation described in Chapter 6 (Equation 6.39). Equation 7.15 forms the boundary condition for Equation 7.3 at the wafer surface. Once the concentration distribution of $R$ is computed the removal rate of copper may be computed.

### 7.2.3 Copper Removal Rate and Effectiveness Factor

Here we use the law of the conservation of mass to develop an expression for the removal rate of copper. Since copper is removed by two mechanisms -- chemical ($r_1$) and mechanical ($r_3$) -- the rate of change of its mass divided by its molecular weight is given by the sum $r_1 + r_3$. Removal rates are usually described by the rate of change of the thickness of the material being removed so that $RR(r,\theta) = \frac{df}{dt}$. The model assumes that copper is removed simultaneously by the following two processes: (i) conversion into $\underline{L}$ followed by abrasion of $\underline{L}$ at the rate $r_1$ ($= r_2$) and (ii) direct abrasion at the rate $r_3$. Thus, from Equations 7.9, 7.11, and 7.13, the instantaneous rate of copper removal at any point ($r$, $\theta$) on the wafer surface is given by

$$\text{RR}(r,\theta) = \frac{MW}{\rho}(r_1 + r_3) = \frac{MW}{\rho} \frac{k_3 + k_1 C(r,\theta,h)}{1 + \frac{k_1}{k_2}C(r,\theta,h)}, \quad (7.16)$$

where MW and $\rho$ denote the molecular weight and density of copper, respectively. Since the wafer rotates about its center and the duration of time over which the wafer is polished is typically much greater than the period of revolution, the average rate of copper removal for points on the wafer at a distance $r$ from its center is given by

$$\text{RR}(r) = \frac{1}{2\pi} \int_0^{2\pi} \frac{MW}{\rho} \left( \frac{k_3 + k_1 C(r,\theta,h)}{1 + \frac{k_1}{k_2}C(r,\theta,h)} \right) d\theta. \quad (7.17)$$

RR(r) is the local removal rate. The average removal rate for the entire wafer surface, $\text{RR}_{avg}$, is obtained by integrating over the area of the wafer to obtain

$$\text{RR}_{avg} = \frac{2}{R_1^2} \int_0^{R_1} \text{RR}(r) r \, dr. \quad (7.18)$$

An effectiveness factor, $\eta$, is defined by

$$\eta = \frac{\text{RR}_{avg}}{\text{RR}_{avg,0}}, \quad (7.19)$$

where $\text{RR}_{avg,0}$ is given by Equation 7.17 and Equation 7.18 with $C(r, \theta, h)$ replaced by $C_0$ and with $k_2$ and $k_3$ replaced by the reference values $k_{20}$ and $k_{30}$, respectively, so that

$$\text{RR}_{avg,0} = \frac{MW}{\rho} \left( \frac{k_{3_0} + k_1 C_0}{1 + \frac{k_1}{k_{2_0}} C_0} \right). \quad (7.20)$$

The average removal rate would take the value of $\text{RR}_{avg,0}$ if the entire wafer surface were exposed to the inlet concentration. This would be the case if mass transport occurred very fast relative to the time scale for the surface kinetics so that the surface concentration would be $C_0$ to a good approximation. The effectiveness factor, $\eta$, is thus a measure of the mass-transport resistance to the overall removal process. That is, $\eta$ is very close to unity if the mass transport rate of $R$ to the surface is fast enough to feed the chemical reaction and maintain the concentration at the slurry-surface interface near the inlet concentration, $C_0$.

### 7.2.4 Kinetic Rate Parameters

The removal mechanism of copper has been modeled using three surface kinetic parameters $k_1$, $k_2$ and $k_3$. The parameter $k_1$, which describes the rate of the chemical reaction at the wafer surface (Equation 7.6), is primarily a function of the oxidizer R in the slurry. The local temperature, pressure and concentration of R are also known to effect the rate of chemical reaction. In CMP, the abrasive particles also can enhance the rate of reaction by weakening the surface bonds by microcutting or brittle fracture. Here, we assume that $k_1$ depends only on the oxidizer used and is independent of position on the wafer surface.

The rate parameters $k_2$ and $k_3$ describe the rates of mechanical abrasion of $\underline{L}$ and $\underline{Cu}$, respectively. The contact pressure, pad, abrasive size and concentration have strong effect on $k_2$ and $k_3$. The value of $k_2$ also depends strongly on the type of reacted layer, $\underline{L}$, formed by the oxidizer. In the case of oxidizers such as potassium dichromate ($K_2Cr_2O_7$), the reacted layer passivates the surface [7.14] and prevents further corrosion of the underlying copper. Other oxidizers, such as nitric acid ($HNO_3$) [7.15] and ferric nitrate ($Fe(NO_3)_3$) [7.16], do not form an effective passivating film. The properties of the reacted layer may depend upon the concentration of the oxidizer in the slurry as in the case of hydrogen peroxide ($H_2O_2$) [7.16]. Lower concentrations of $H_2O_2$ give a porous reacted layer while higher concentrations produce a passivating layer. Note that $k_3$ is the rate at which copper is abraded and is not a function of the slurry oxidizer.

The values of the abrasion rate parameters $k_2$ and $k_3$ vary locally as a function of the local contact pressure at the wafer surface. To illustrate the effect of variation of the contact pressure on the local removal rate, we assume that $k_2$ and $k_3$ are directly proportional to the local contact pressure between the wafer and pad in some simulation runs. The mechanical rate parameters $k_2$ and $k_3$ are assumed to be proportional to a shape function $\phi(r)$, which describes the shape of the contact pressure distribution,

$$k_2(r) = k_{20}\phi(r), \qquad k_3(r) = k_{30}\phi(r). \tag{7.21}$$

The shape function captures the positional dependency of $k_2$ and $k_3$ across the wafer, while the proportionality parameters $k_{20}$ and $k_{30}$ (which are functions of applied pressure, wafer/pad speed, abrasives, back pressure, wafer-backing film and the type of pad) determine the scale.

Several articles have been published which present models to compute the contact pressure distribution [7.17-7.21]. Baker's 2-D contact model [7.17], for example, treats the pad as a beam supported by an elastic foundation; this model leads to an analytical expression for the shape, $\phi(r)$, of the contact pressure distribution which takes the form

$$\phi(r) = 1 + \{1 + 4\cos(-\beta(R_1 - r))\}\exp(-\beta(R_1 - r)), \tag{7.22}$$

where $\beta$ is a parameter which depends on the elastic properties of the pad. The shape function, $\phi(r)$, near the wafer edge is plotted in Figure 7.4 for a typical value of $\beta$ given by Baker [7.17]. The contact pressure is constant in the central region of the wafer and there is a six-fold increase at the edge. The parameter $\beta$ primarily determines the distance from the edge at which the contact pressure starts increasing. For the choice of $\beta$ used here, the contact pressure starts increasing approximately 4 mm away from the edge.

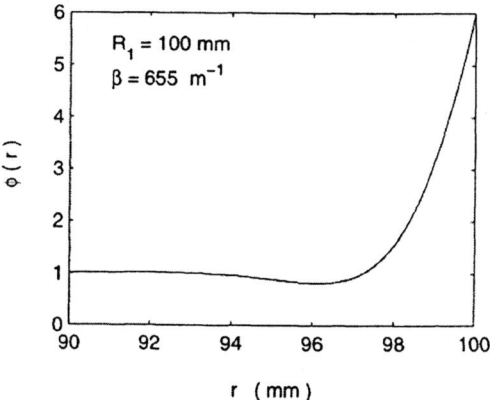

*Figure 7.4:* Shape function for the contact pressure distribution (plotted near the wafer edge) from Baker's model [7.17].

### 7.2.5 Solution Procedure

A numerical method of solution is needed to obtain the spatial distribution of oxidizer concentration, $C$, in the region between the wafer and the pad. Once $C$ is computed, the copper removal rate is obtained using Equation 7.16. Equation 7.3 is solved numerically for $C(r, \theta, z)$ in the region between the wafer and pad defined by $r \in [0, R_1]$, $\theta \in [0, 2\pi]$ and $z \in [0, h]$, with boundary conditions at inlet, pad and wafer surfaces given by Equations. 7.4, 7.5 and 7.15, respectively.

Since the wafer surface is curved, transformation of the $z$-coordinate is convenient for the purpose of discretization. The following coordinate transformation is applied

$$\varsigma = z / h(r, \theta), \qquad (7.23)$$

which changes the solution domain from $z \in [0, h]$ to $\varsigma \in [0, 1]$. Equation 7.3 and its boundary conditions are changed to the $(r, \theta, \varsigma)$ coordinate system and then discretized using a finite-volume approach. A method of iteration (Newton's method) is used to solve the system of discrete equations for the oxidizer concentration. The accuracy of the numerical results were checked by using standard grid refinement techniques [7.2]. For the results reported in section 7.3, the number of grid points in the $r$-, $\theta$- and $\varsigma$-directions are 75, 50 and 50, respectively, for a 200 mm diameter wafer.

## 7.3 MODEL RESULTS

As described in the previous section, the spatial distribution of oxidizer concentration, $C(r,\phi,z)$, and the local removal rate of copper, $RR(r)$, may be computed once the flow and mass-transport problems are solved. Three sample run conditions and key results are presented in Table 7.1, with more results presented elsewhere [7.1, 7.2]. A typical mass diffusivity of a species in liquid media of $10^{-9}$ m$^2$s$^{-1}$ is used along with a stoichiometric coefficient, n, of unity. Here, n=1 represents the case where one molecule of reactant reacts with one molecule of copper, in contrast to the n=12 case used for a densely crosslinked polymer unit (see section 6.4.3). The slurry flow is used as input to the mass-transport equations, and the direct abrasion rate coefficient, $k_{30}$, is set to zero since a desirable CMP process is dominated by the removal of the oxidized surface layer, i.e. $r_1 = r_2 \gg r_3$. The results listed in Table 7.1 show that the average removal rate decreases with decreasing $k_1$, a measure of the rate of chemical reaction at the wafer surface. The values of $k_1$ are varied by a factor of more than 20 to show how the chemical reaction affects the results. The behavior of $\eta$ and the average concentrations of the oxidizer at the wafer and pad, denoted by $C_s$ and $C_b$, respectively, as functions of $k_1$ are also indicated in the Table 7.1.

Figure 7.5(a) shows contours of the normalized concentration distribution, $C/C_0$, of the oxidizer at the wafer surface ($\zeta = 1$) for Case 1. Similarly, Figures 7.5(b) and 7.5(c) show plots of $C/C_0$ contours on surfaces $\zeta = 0.75$ and $\zeta = 0.5$, respectively, between the wafer and pad. The value of $C/C_0$ is equal to 1 at the slurry inlet regions (the bottom parts of the wafer edge in the plots) as set by the boundary condition at the inlet. The oxidizer reacts with the copper at the wafer surface and is thus consumed during flow underneath the wafer.

*Table 7.1:* Summary of results of the three different cases. Simulation parameters: $C_0 = 0.1$ mol/lit, $D = 10^{-9}$ m$^2$s$^{-1}$, $k_{20} = 1 \times 10^{-2}$ mol m$^{-2}$s$^{-1}$, $k_{30} = 0$, $\beta = 655$ m$^{-1}$ and n = 1. Flow simulation parameters are the same as in Figure 7.3.

| Case | $K_1 \times 10^5$ (m/s) | $RR_{avg}$ (nm/min) | $\eta$ | $C_s/C_0$ | $C_b/C_0$ |
|---|---|---|---|---|---|
| 1 | 15 | 1256 | 0.5 | 0.3 | 0.85 |
| 2 | 3 | 677 | 0.7 | 0.65 | 0.93 |
| 3 | 0.6 | 215 | 0.9 | 0.9 | 0.97 |

# COPPER CMP MODEL BASED UPON FLUID MECHANICS AND SURFACE KINETICS

The depletion of the oxidizer is higher near the wafer surface as it is consumed by the surface reaction, as illustrated by the contours of $C/C_0$ in the $y$-$z$ coordinate plane shown in Figure 7.5(d). (The slurry flows from left to right in this plot.) The value of the concentration near the wafer surface is approximately $0.3C_0$ except for a small region near the inlet, while the concentration near the pad is almost undepleted with values ranging from approximately 0.75 to 1.

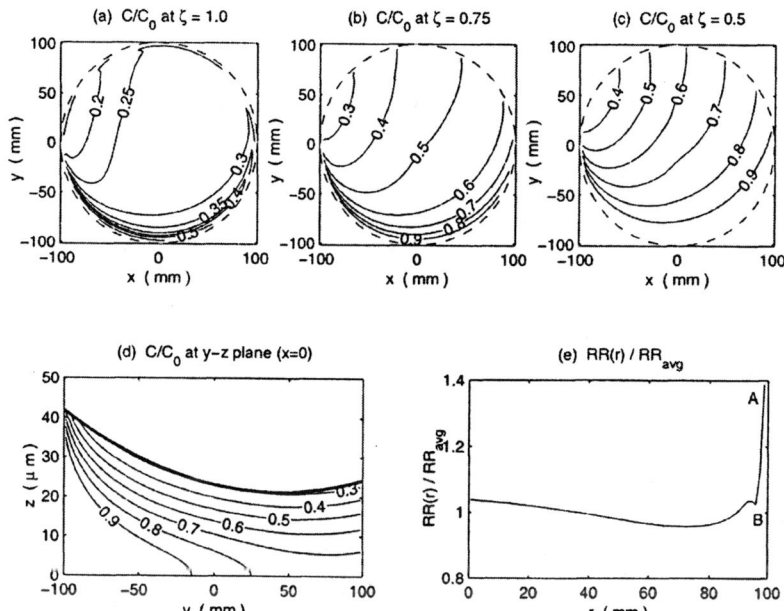

Figure 7.5. Case 1: (a-c) Contours of normalized concentration distribution ($C/C_0$) of reactant, R, at different $\zeta$ = constant surfaces between the wafer (at $\zeta = 1$) and pad (at $\zeta = 0$). Wafer edge is shown by the dashed line. (d) $C/C_0$ contours of R at the $y$-$z$ coordinate plane. Wafer surface is shown by the thick line. (e) Copper removal rate as a function of radial position. $k_1 = 15 \times 10^{-5}$ m/s. Other simulation parameters are listed in Table 7.1. Ref [7.1].

Figure 7.5(e) shows the normalized local copper removal rate as a function of wafer radial position, $r$, that is the local removal rate $RR(r)$ divided by the average removal rate $RR_{avg}$. The computed average removal rate, $RR_{avg}$, is 1256 nm/min for Case 1. The high removal rate at the edge (labeled A in the figure) is due to the large value of $k_2$ (see Equation 7.21) at the edge. The small dip at point B corresponds to the dip in the value of $k_2$ at $r = 96$ mm as seen in Figure 7.4. (The change in slope of $RR(r)$ near B is very rapid and is not a slope discontinuity as it appears in the plot.) The high removal rate consumes more oxidizer near the edge, contributing to the decrease in the removal rate towards the left of B.

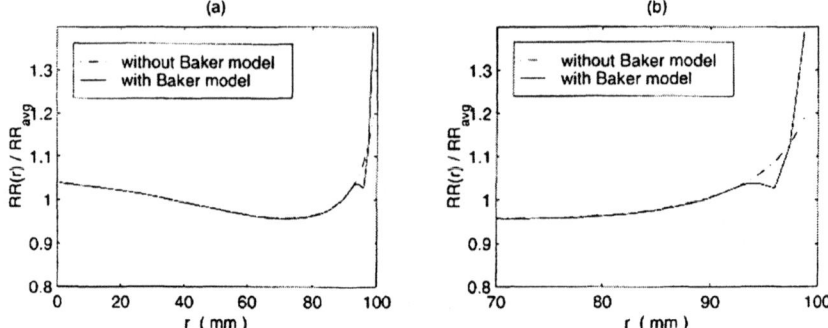

*Figure 7.6:* (a) Normalized local removal rate variation with and without Baker's model [7.17]. (b) Enlarged view near the edge showing the difference. Ref [7.1].

The effect of Baker's contact model on the radial variation in the removal rate is shown in Figure 7.6. The solid curve in the plots is Case 1 which uses $\phi(r)$ from Equation 7.22 (Baker model), while the dashed curve is obtained using no contact model, i.e. $\phi(r)=1$. The removal rate variation in the central region is purely due to effects of mass transport and hydrodynamics with Baker's model only affecting the results near the edge (enlarged in Figure 7.6 (b)). Cases 2 and 3 have lower values of $k_1$ as compared to Case 1 (see Table 7.1). As $k_1$ decreases the rate of chemical reaction decreases at the wafer surface, with a corresponding decrease in removal rates as listed in Table 7.1. A higher value of $k_1$ results in a lower average surface concentration of the oxidizer, $C_s$, indicating (i) a fast wafer surface reaction rate and (ii) that mass transport of oxidizer to the wafer surface plays a role in the removal process of copper. Table 7.1 also lists the average concentration of the oxidizer at the pad surface $C_b$, which does not vary as much as $C_s$ because it is further away from the reacting surface and in effect is the source from which diffusion occurs to feed the reaction.

The impact of mass transport on the removal rate is quantified by the efficiency factor, $\eta$, as defined in Equation 7.19 which varies from 0 to 1. The efficiency factor for Case 1 is lowest indicating that mass-transport resistance of the oxidizer in reaching the wafer surface plays an important role in the removal process. In Case 3, a high value of $\eta = 0.9$ indicates that mass transport plays a less significant role as compared to Cases 1 and 2.

Figure 7.7 shows the effect of varying $k_1$, $k_{20}$, $k_{30}$ and $C_0$ on the average removal rate. Values for the other input parameters are fixed and are taken to be $D = 10^{-9}$ m$^2$s$^{-1}$, $\beta = 655$ m$^{-1}$ and n = 1/3, and the remaining flow simulation parameters are the same as those used previously. Figure 7.7(a)

# COPPER CMP MODEL BASED UPON FLUID MECHANICS AND SURFACE KINETICS

indicates that the average removal rate increases steadily with $C_0$ at small values of $C_0$ and then plateaus to a maximum at high values of $C_0$. At small values of $C_0$, the removal rate is limited by the rate of chemical reaction (Equation 7.6), whereas at larger values of $C_0$ it is limited by the rate of mechanical abrasion (Equation 7.8). The chemical rate parameter, $k_1$ determines the slope of the curve of $RR_{avg}$ versus $C_0$ at small values of $C_0$, while the mechanical rate parameter, $k_{20}$, determines the height of the plateau at larger values.

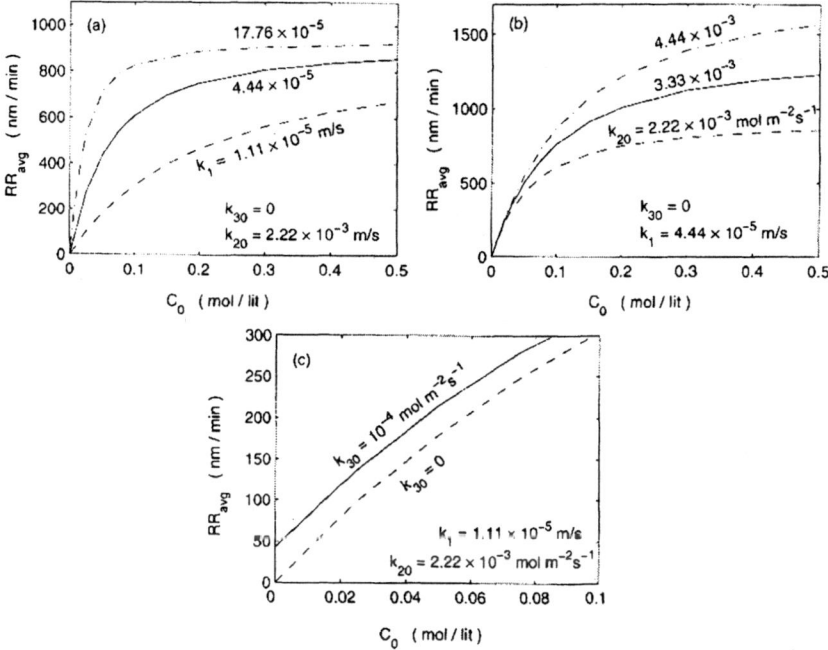

*Figure 7.7:* Behavior of the average removal rate as a function of the inlet concentration of the oxidizer for (a) varying $k_1$, (b) varying $k_{20}$, and (c) varying $k_{30}$. Ref [7.1].

In Figure 7.7(a), the three curves shown have different initial slopes which are directly proportional to their respective values of $k_1$. All three curves approach the same maximum value of $RR_{avg}$ at large values of $C_0$ (not shown in the figure) since $k_{20}$ is the same for all the curves. In Figure 7.7(b), the removal rate curves have the same initial slope and reach different plateaus heights which are directly proportional to their respective $k_{20}$ values. Figure 7.7(c) shows the direct abrasion of copper (i.e. parameter $k_{30}$) causes an approximately uniform increment in removal rate at all oxidizer concentrations. Direct abrasion of copper is the only mechanism of removal in our model when the slurry has zero oxidizer concentration ($C_0=0$), so that $k_{30}$ can be determined experimentally using a slurry with $C_0 = 0$.

## 7.4 COPPER CMP EXPERIMENTS WITH POTASSIUM DICHROMATE-BASED SLURRY

Copper CMP experiments were performed with a $K_2Cr_2O_7$ slurry, with the applied pressure, wafer/pad speed and oxidizer concentration varied. The objectives were to (i) evaluate the model assumptions, (ii) determine values of the kinetic parameters $k_1$, $k_{20}$ and $k_{30}$ by fitting the model predicted average removal rate to the experimentally measured value, and (iii) determine the behavior of $k_{20}$ and $k_{30}$ as functions of applied pressure and speed.

Blanket copper films were sputtered on 125 mm diameter unpatterned wafers. The slurry consisted of an oxidizer ($K_2Cr_2O_7$), alumina particles (50 nm nominal diameter) and de-ionized (DI) water, with the abrasive concentration fixed at 3 wt%, as originally investigated by B.-C. Lee [7.22-7.23]. An IPEC 372M polishing tool was used with a Pan-W pad manufactured by Freudenberg. The diameter of the polishing table was 57 cm and the eccentric distance between the polishing head and the center of the polishing table, $R_2$, was 173 mm. Both the platen and carrier were rotated at equal speed in all experiments, with the slurry delivery rate fixed at 150 ml/min. The polish time varied from 1-2 min. Sheet resistance was measured using a four-point probe before and after the polishing experiment at four locations in the central region of the wafer. The sheet resistance measurements were used to obtain the local removal rates, which were then averaged to compute an average removal rate, $RR_{avg}$.

The $K_2Cr_2O_7$ slurry is ideal to evaluate our model because the formation of a reacted layer on the surface with this slurry has been confirmed by potentiodynamic and surface Auger electron spectroscopy (AES) measurements [7.14]. The overall oxidation reaction at the wafer surface can be written as follows:

$$3Cu + Cr_2O_7^{2-} + 14H^+ \rightarrow 3Cu^{2+} + 2Cr^{3+} + 7H_2O \qquad (7.27)$$

Three moles of copper react per mole of the oxidizer, giving a stoichiometric coefficient n=1/3. (For comparison with Equation 7.2, R is $Cr_2O_7^{2-}$ and L is $Cu^{2+}$.) The applied pressure, wafer/pad speed and $K_2Cr_2O_7$ concentration in the slurry were varied to study their effects on the average removal rate. The following pressure(kPa)/speed(rpm) combinations were used: 14/45, 28/45, 42/45, 28/30 and 28/60. $K_2Cr_2O_7$ concentration in the slurry, $C_0$, was varied between 0 and 0.35 mol/lit for each pressure/speed combination. The experimental data plotted in Figure 7.8 shows that the average removal rate increases with increase in either pressure or speed. A measurable removal rate exists at zero $K_2Cr_2O_7$ concentration which is also function of pressure and speed. For any given pressure/speed combination, $RR_{avg}$ initially increases with $C_0$ and then plateaus with further increase in $C_0$, in agreement with our model.

The values of $k_1$, $k_{20}$ and $k_{30}$ for this slurry are determined by fitting the average removal rate predicted by our model with that determined experimentally. We assume the following: (i) the value of $k_1$ is constant and (ii) $k_{20}$ and $k_{30}$ are functions of pressure and speed. The shape function, $\phi$, is set equal to one as it only affects the removal rate in a small region at the wafer edge and does not significantly affect the average removal rate calculations. The value of $k_1$, which essentially determines the rate of change of $RR_{avg}$ with $C_0$ for small values of $C_0$, is estimated by fitting to the initial slopes at all pressure/speed combinations of the experimental data shown in Figure 7.8. At a given pressure and speed, $k_{20}$ is estimated to fit the removal rate data in the plateau region, while $k_{30}$ is estimated to fit the data at zero oxidizer concentration. The efficiency factor, $\eta$, for all experimental conditions is very close to 1 implying that sufficiently high pressures and rotation rates have been used to have rapid mass transport of the oxidizer. Step 1 does not play a significant role in the copper removal process for this slurry. The removal process is controlled by the surface kinetic steps, and thus the average removal rate is well approximated by Equation 7.20.

The six sets of data plotted in Figure 7.8 correspond to the five chosen pressure/speed combinations in the experiments. Given that $\eta$ is close to 1, values for $k_1$, $k_{20}$ and $k_{30}$ may be determined from the data using a least squares fit to the curves given by Equation 7.20 with the constraint that $k_1$ is held fixed for all data sets [7.1, 7.2]. The fitted rate parameters are listed in Table 7.2.

It is evident that $k_{20}$ and $k_{30}$ are increased by increasing either $P_{app}$ or the rate of rotation. This occurs because both of these rate constraints, $k_{20}$ and $k_{30}$, reflect mechanically assisted removal (desorption) of either an altered layer or a copper surface. $P_{app}$ and the rate of rotation also affect the fluid mechanics, which in turn can increase or decrease the rate of mass transfer. Consequently, the input parameters $P_{app}$ and $\omega$ affect all aspects of the removal mechanism.

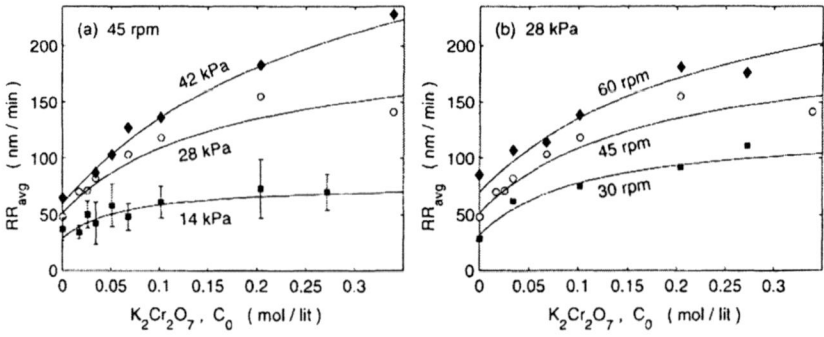

*Figure 7.8.* Average copper removal rate as a function of $K_2Cr_2O_7$ concentration for (a) varying applied pressure and (b) varying speed. The error bars for the 45 rpm/14 kPa case indicate one standard deviation in the average removal rate. The solid curves show model predictions based on fitted values of the kinetic parameters listed in Table 7.2. Ref [7.1].

In the experimental range of pressure and speed, both $k_{20}$ and $k_{30}$ are approximately proportional to the wafer-pad speed while $k_{20}$ has a stronger dependency on the applied pressure than that for $k_{30}$. Intuitively, the hardness of the copper layer relative to that of the reacted layer plays a significant role on the dependency of $k_{20}$ and $k_{30}$ on pressure, but not so much on speed. For the oxidizer used in the experiments, the dependency suggests that the copper layer is harder than the reacted layer. Also, at a given pressure and speed, the higher values of $k_{20}$ as compared to $k_{30}$ (both fitted from the experiments) again implies the reacted layer is easier to abrade than copper.

*Table 7.2.* Fitted values of $k_1$, $k_{20}$, and $k_{30}$ for different pressure/speed combinations.

| $P_{app}$ (kPa) | $\omega_1 = \omega_2$ (rpm) | $k_1 \times 10^6$ (ms$^{-1}$) | $k_{20} \times 10^4$ (mol m$^{-2}$s$^{-1}$) | $k_{30} \times 10^4$ (mol m$^{-2}$s$^{-1}$) |
|---|---|---|---|---|
| 14 | 45 | 2.82 | 1.83 | 0.68 |
| 28 | 45 | 2.82 | 4.88 | 1.21 |
| 42 | 45 | 2.82 | 8.61 | 1.40 |
| 28 | 30 | 2.82 | 2.96 | 0.75 |
| 28 | 60 | 2.82 | 6.95 | 1.64 |

Removal by chemical reaction followed by the abrasion of the reacted layer is preferred as a better surface finish is obtained. Wafers which were polished with $C_0 > 0.1$ mol/lit had a shiny metallic finish, while polishing with $C_0 < 0.1$ mol/lit resulted in a hazy finish with significant visible scratching. This observation is aligned with previous observations with SiLK polymer (Figure 5.16).

## 7.5 SUMMARY

A model for copper CMP has been described here, based on the model developed by Thakurta et al. [7.1] for an acidic slurry. The model takes into account the effects of slurry hydrodynamics, slurry chemistry, chemical reaction and mechanical abrasion at the wafer surface. The steps in the copper removal process are similar to those described in Chapter 6 for low-κ CMP, and include: mass transport of the oxidizer to the wafer surface, reaction of oxidizer with copper to form a reacted layer with subsequent removal of the reacted layer by mechanical abrasion; in addition copper may be removed by direct abrasion. The rates of the chemical reaction and mechanical abrasion steps are described by separate kinetic rate parameters. Variation in the removal rate results from mass transport effects and variation in the local contact pressure.

The Cu CMP experimental results further indicate that a good CMP process requires a balance between chemical and mechanical effects, in agreement with the results for polymer CMP (Chapter 4) and organosilicate CMP (Chapter 5). For low oxidizer concentrations, the chemical reaction (Step 1) limits the removal rate; as a result, surface damage is possible due to direct mechanical abrasion (Step 2) and the removal rate is lower than can be achieved. At high oxidizer concentrations, mechanical abrasion controls the removal rate of the chemically-altered surface layer. Copper CMP experiments with $K_2Cr_2O_7$ oxidizer and alumina abrasive particles are compatible with the proposed surface kinetics model.

The importance of mass transport has been incorporated in our model and is measured by the efficiency factor, $\eta$. Mass transport effects are important when the removal rates are quite high; more specifically, mass transport of the oxidizer to the wafer surface not only limits the removal rate, but results in a wafer-scale radial non-uniformity. Unfortunately, this regime of operation could not be experimentally evaluated with our equipment and consumable set. In fact, any robust CMP manufacturing process requires an $\eta=1$ to avoid wafer-scale non-uniformities in removal rate.

## 7.6 REFERENCES

[7.1] D.G. Thakurta, D.W. Schwendeman, R.J. Gutmann, S. Shankar, L. Jiang and W.N. Gill, Thin Solid Films, accepted for publication (2002).
[7.2] D.G. Thakurta, Ph.D. Thesis, Rensselaer Polytechnic Institute, Troy, NY (2001).
[7.3] S. R. Runnels and L. M. Eyman, J. Electrochem. Soc., **141**, 1698 (1994).
[7.4] D. G. Thakurta, C. L. Borst, D. W. Schwendeman, R. J. Gutmann, and W. N. Gill, Thin Solid Films, **366**, 181 (2000).

[7.5] T. -K. Yu, C. C. Yu, and M. Orlowski, in IEDM Tech. Dig., 865 (1993).
[7.6] M. Bhushan, R. Rouse, and J. E. Lukens, *J. Electrochem. Soc.*, **142** 3845 (1995).
[7.7] D. Stein, D. Hetherington, M. Dugger, and T. Stout, *J. Electronic Materials*, **25**, 1623 (1996).
[7.8] L. Jiang and S. Shankar, *Proc. of 16th Intl. VLSI Multilevel Interconnection Conference (VMIC)*, 245 (1999).
[7.9] N. Patir and H. S. Cheng, *ASME J. Lubrication Tech.*, **100**, 12 (1978).
[7.10] D. G. Thakurta, C. L. Borst, D. W. Schwendeman, R. J. Gutmann, and W. N. Gill, *J. Electrochem. Soc.*, **148**(4), G207-214 (2001).
[7.11] S. Sundararajan, D. G. Thakurta, D. W. Schwendeman, S. P. Murarka, and W. N. Gill, *J. Electrochem. Soc.*, **146**, 761 (1999).
[7.12] R. S. Subramanian, L. Zhang, and S. V. Babu, *J. Electrochem. Soc.*, **146**, 4263 (1999).
[7.13] J. M. Smith, *Chemical Engineering Kinetics*, McGraw-Hill, New York (1970).
[7.14] R. J. Gutmann, C. L. Borst, B. -C. Lee, D. Thakurta, D. J. Duquette, and W. N. Gill, *Proc. of 17$^{th}$ Intl. VLSI Multilevel Interconnection Conference (VMIC)*, Santa Clara, CA, 123 (2000).
[7.15] J. M. Steigerwald, Ph. D. Thesis, Rensselaer Polytechnic Institute, Troy, NY (1995).
[7.16] Q. Luo, S. Ramarajan, and S. V. Babu, *Thin Solid Films*, **335**, 160 (1998).
[7.17] A. R. Baker, *Electrochem. Soc. Fall Meeting, extended abstracts (EA 96-2)*, San Antonio, TX (1996).
[7.18] C. Srinivasa-Murthy, D. Wang, S. P. Beaudoin, T. Bibby, K. Holland, and T. Cale, *Thin Solid Films*, **533**, 308 (1997).
[7.19] D. Wang, J. Lee, K. Holland, T. Bibby, S. Beaudoin, and T. Cale, *J. Electrochem. Soc.*, **144**, 1121 (1997).
[7.20] D. Wang, A. Zutshi, T. Bibby, S. Beaudoin, and T. Cale, Effects of carrier film physical properties on W CMP, *Thin Solid Films*, **345**, 278 (1999).
[7.21] J. Tichy, J. A. Levert, L. Shan, and S. Danyluk, *Electrochem. Soc.*, **146**, 1523 (1999).
[7.22] B.-C. Lee, Ph.D. Thesis, Rensselaer Polytechnic Institute, Troy, NY (2000).
[7.23] B.-C. Lee, B. Wang, D. J. Duquette, R. J. Gutmann, *Proc. of 6$^{th}$ Intl. CMP for Multilevel Interconnect Conf. (CMP-MIC)*, Santa Clara, CA, 47 (2000).

Chapter 8

# FUTURE DIRECTIONS IN IC INTERCONNECTS AND RELATED LOW-κ ILD PLANARIZATION ISSUES

Fully planarized advanced interconnect structures with copper metallization and ultra low-κ ILDs fabricated by dual damascene patterning currently prevent conventional ICs from being interconnect limited. However, as CMOS scaling and lithography advances reduce the minimum feature size below 50 nm, interconnects will again become a performance limiter and probably a manufacturing cost enhancer. Unlike the situation in the late 80s, new materials and a new patterning process used with an increasing amount of interconnect levels will not be a viable solution. After copper metallization with atomic-scale liners and dielectrics with κ<1.8, conventional approaches are not compatible with the performance needed with sub-50nm devices.

Various technologies have been proposed and are being investigated as a post-copper/low-κ interconnect approach. These include optical interconnects, multiplexed interconnects, and three-dimensional (3D) integration. However, all these approaches require aggressive conventional IC interconnect structures as basic building blocks. More specifically, only the upper (global) interconnect levels will be affected; the lower levels will remain as conventional scaled interconnects. Therefore, this chapter begins with a discussion of CMP-related issues with ultra low-κ (i.e., porous) dielectrics and alternative planarization techniques being pursued. This section is followed by a discussion of alternative interconnect technologies to address post-50 nm long-range interconnect limitations, and concludes with a more detailed discussion of 3D integration -- the most promising of the non-conventional alternatives. Issues with low-κ planarization are highlighted throughout.

## 8.1 PLANARIZATION OF INTERCONNECTS WITH ULTRA LOW-κ ILDS

Damascene patterning of copper interconnects with ultra low-κ (κ<1.8) porous dielectrics is a formidable research, development and manufacturing challenge, particularly at a minimum feature size (MFS) ~ 50nm earmarked in the ITRS. Major challenges include reactive ion etching of high-aspect-ratio (HAR) vias and trenches, liner and copper fill, and planarization back to the dielectric surface. The challenge is to leave a planar structure with embedded copper interconnects and a damage-free dielectric for subsequent processing of additional interconnect levels. Since CMP involves abrasive particles of greater dimension than the MFS of future interest and requires down pressure with concerns of ILD mechanical stability, alternatives to conventional CMP are being considered. In this section the most promising alternatives to date are summarized, followed by predictions of future planarization trends.

Planarization alternatives to CMP are grouped as follows:
1) CMP based approaches: alternatives to conventional CMP with abrasive particles and rotary tools (including linear belt and indexed-pad tools as well as pads with embedded particles)
2) non-CMP based approaches: reactive liquid planarization, spin etch planarization, and electropolishing. In each of these approaches, abrasive particles are not involved in a mechanical removal process, so appreciably less mechanical damage is expected. The planarization efficiency of these processes has not been established.

### 8.1.1 Alternatives to Conventional CMP

As described in Chapter 3, conventional CMP uses a rotary platform with pad asperities (and often grooves) to distribute a slurry with abrasive particles to the wafer surface. The slurry chemically modifies the surface of the material to be removed and carries away the abraded material. Various alternatives have been and/or are being pursued to improve the planarization process. Improvements in this sense include increased removal rate with improved within wafer and wafer-to-wafer uniformity, less down-force and shear stress to minimize sub-surface damage and adhesion issues (particularly important with low-κ and ultra low-κ dielectrics), and less surface defects after post-CMP cleaning.

Many alternative CMP tools have been developed:
1) orbital tools with the slurry mixed near point-of-use and delivered through the pad (IPEC, now Speedfam)

# FUTURE DIRECTIONS IN IC INTERCONNECTS AND RELATED LOW-K ILD PLANARIZATION ISSUES

2) linear tools in which a continuous pad/belt is used to replace rotary platen motion (using water or air to maintain contact with the wafer and to maintain removal-rate uniformity) (OnTrak/Lam)
3) web-based indexed pads to maintain process uniformity with less process down time (Obsedian, now Applied Materials)

In most cases, tool reliability, metrology and post-CMP cleaning appear to dominate in the market. Besides the use of abrasive particles like silica, alumina and ceria, use of fixed abrasive pads (initiated by 3M and licensed to Rodel [8.1-8.3]) may been a promising alternative to conventional CMP. With a fixed abrasive contained in the surface region of the pad, only chemical liquid needs to be distributed to the CMP tool. Such a process is preferred within an IC manufacturing facility where both slurry particle distribution to the CMP tool and removal in the waste stream are concerns. However the fixed abrasive pads are not easy to modify as CMP process requirements evolve. As a result, this approach has not been used regularly in manufacturing environments to date.

## 8.1.2 Non-CMP Approaches to Planarization

While conventional CMP is well established, the copper damascene patterning process is particularly demanding and needs to be done 6-8 times in advanced ICs. When we consider the present move to low-$\kappa$ ILDs and the future move to ultra low-$\kappa$ ILDs by 2008, the desire to replace conventional CMP is apparent. Removal of the excess copper in the field regions after copper fill needs to be done in a mechanically benign fashion to alleviate mechanical stress on the relatively fragile ultra low-$\kappa$ materials. Three approaches being explored are the use of reactive liquids, spin-etch planarization and electropolishing.

Reactive liquids are used with a conventional CMP tool, but without abrasive particles or fixed abrasive pads. A liquid, which is more chemically active than a normal CMP slurry, is used. Planarization occurs because the pressure applied by the pad to the high features enhances the removal rate compared to the removal in the low areas [8.4-8.5]. The removal rate versus pressure curve assumes an S-shape (compared to the linear dependence in conventional CMP following Preston's Equation).

Reactive liquids have demonstrated high removal rates, but the planarization effectiveness to date is less than with conventional CMP. There is less fundamental knowledge to date with this approach, but a low-abrasive concentration in a conventional CMP slurry combined with some reactive liquid chemistry seems to incorporate many of the advantages of "abrasive-free" while minimizing the impact of CMP slurry distribution issues described previously. However, it is possible that non-conventional pads could be developed that would enhance the planarization capability of reactive liquids.

Spin-etch planarization involves spinning a wafer on a contamination-free nitrogen cushion while delivering fluids across the wafer surface in a controllable fashion [8.6-8.7]. As with reactive liquid CMP, no abrasive particles are used. While such a technology platform has been demonstrated for backside-substrate removal of metallic contaminants, the applicability for DD patterning (and planarization in general) is unproven. As with reactive liquid-based CMP, this process is low stress (and therefore compatible with ultra low-κ ILDs), but with unproven planarization and DD process compatibility.

Electropolishing is a classical technique for cleaning and polishing of copper. Success has been reported in achieving high removal rates with blanket copper films [8.8] using well established approaches [8.9]. Recent work indicates strong potential for a stress free process [8.10], although the planarization potential has yet to be established.

One of the more novel techniques for low-stress copper planarization is electrochemical mechanical deposition (ECMD). ECMD uses a porous membrane in close proximity to the wafer surface during electroplating. The effect is simultaneous copper deposition and planarization, as the copper plating is more favorable within features, and the plated overburden is immediately "polished", or brushed away from the field areas [8.11]. The result is a very planar copper surface with reduced overburden and near-zero topography (as shown in Figure 8.1). This method may enable spin-etch planarization for Cu removal in DD patterned structures, but the copper fill/void properties and defect density are not yet proven.

*Figure 8.1* Dual damascene trenches of varied linewidth plated using ECMD with greatly reduced overburden and near-zero topography.

The use of low stress planarization becomes critical with ultra low-κ ILDs and/or when air gaps or air bridges are used to reduce coupling capacitance to a minimum. Air bridges common in microwave monolithic integrated circuits (MMICs) have only recently been proposed for silicon ICs with a large number of interconnect levels [8.12, 8.13]. If adapted in the future, the low-stress planarization alternatives briefly summarized here will be increasingly required. However, conventional CMP is now an established IC process which can meet stringent processing requirements and will prove difficult to replace.

## 8.2 ALTERNATIVES FOR THE POST-COPPER/ULTRA LOW-κ INTERCONNECT ERA

While the back-end-of-the-line (BEOL) technological advances of the 90s described in Chapter 1 appreciably reduced the interconnect bottleneck to continued IC scaling, copper/ultra low-κ interconnect structures become an appreciable performance limiter as the MFS goes below 50nm. Several novel interconnect alternatives and new approaches are highly desirable. These alternative solutions can be grouped as follows:
1) alternative conducting materials (e.g. silver or room temperature superconductors) or dielectric materials (concepts such as κ<1)
2) alternative distribution of signals and clocks (e.g., wireless distribution or multiplexing)
3) non-electrical distribution of signals and clocks (e.g., optical or terahertz)
4) non-planar integrated assemblies (e.g., three-dimensional (3D) integration)

In this section these alternatives are summarized, with the nearest-term, most-viable approach (3D integration) discussed in more detail in Section 8.3. Full-wafer planarization is a demanding process in many wafer-scale 3D implementations, with mechanically fragile ultra-low-κ materials a major impediment to implementation.

### 8.2.1 Alternative Materials

There are no clear cut materials options after copper/ultra low-κ. However, various alternatives are being pursued and investigated to some degree which warrant discussion. Silver has been proposed to replace copper because of a slightly higher electrical conductivity [8.14]. However, the packaging community has a significant database on corrosion and silver migration issues which are disconcerting. While the slight electrical conductivity advantage of silver does not warrant the research and development expense, a major detractor is that silver has a larger mean free path than copper [8.15]. As a result, as the feature size of conducting lines and vias decreases below 100 nm, the effective conductivity of silver degrades more rapidly than copper. Issues of low-κ film adhesion, barriers for silver high-field diffusion, and electromigration have not been adequately addressed to establish long term viability. The authors believe that the larger mean free path (and therefore, more interface scattering) is a key disadvantage of silver, and doubt that silver will receive the necessary scientific and technological investment to challenge copper.

It should be noted that the mean free path of aluminum is less than that of copper, leading to some projections that aluminum may be revisited as the MFS approaches 50nm [8.16]. However, the technology investment and manufacturing experience gained with copper and low-κ dielectrics make such a transition unlikely.

The family of high temperature superconductors provides another long-term conductor alternative. However, even if room temperature superconductors are realized, issues of ambient operating temperature, current handling capability and signal and clock transmission properties need to be evaluated, as well as development of a baseline interconnect process flow. Certainly this is not a viable alternative for ITRS planning purposes, requiring fundamental materials breakthroughs of Nobel prize magnitude.

Recently researchers have proposed the concept of dielectric materials with κ<1, based upon non-conventional concepts of polarization. While index of refraction less than unity have been described for optical materials and configurations, these generally have narrow wavelength/frequency bands of operation. For a viable interconnect technology using baseband distribution of clock and signal lines, a material/structure with broadband (DC to a few GHz) low-κ is required. Such a dielectric requires a material/structure breakthrough as significant as a room-temperature superconductor, or even more so. While perhaps not fundamentally impossible, such a material would require a new understanding of dielectric polarization.

### 8.2.2 Alternative Distribution for Signals and Clocks

As Cu/low-κ interconnects become performance limiters when the MFS is ~50 nm, a new design/interconnect paradigm is required. Both the wiring complexity and interconnect bandwidth will become insufficient, requiring new architectures and/or new interconnect technology. Multiplexing techniques well-established for generations of telecommunications applications and under intense research and development presently for wireless communications can alleviate both wiring complexity and bandwidth constraints. In addition, information can be distributed by either guided wave/transmission line structures or by free-space radiation. The most viable alternatives to alleviate future Cu/low-κ interconnect bottlenecks are summarized in this section.

Any multiplexing technique utilizes a carrier frequency well above the information bandwidth of the signals to be multiplexed. Since information bandwidths will be approaching 5-10 GHz in the time scale envisioned, either millimeter wave, terahertz or optical carrier frequencies/wavelengths will be needed. Multiplexing signal design will be leveraged from ongoing research in wireless communication networks, particularly cellular systems and/or local area networks (LANs); while time-division-multiplexing (TDM) may be suitable in applications where many interconnect paths have low duty cycle operation, spread spectrum techniques such as code-division-multiple-access (CDMA) will be more widely utilized in applications such as generation three (G3) wireless systems.

The choice between millimeter wave and optical techniques for such implementations is not clear; research and development in both areas is envisioned for the foreseeable future. Both alternatives can be used with planar guided wave signal transmission as well as with radiative signal distribution, although the frequency/wavelength difference (~ 3 orders of magnitude) results in a different technology base. Optical has a major advantage in operating bandwidth, while millimeter wave has an advantage in conversion efficiency and power consumption [8.17].

Two innovative approaches using RF/microwave/millimeter techniques to alleviate interconnect bottlenecks are high frequency clock distribution by radiation [8.18-8.19] and CDMA (or possible frequency division multiple access (FDMA)) distribution of signals in a LAN-based architecture using a planar multichip module (MCM) with evanescent wave near-field coupling [8.20]. The first offers the possibility of a lower power clock distribution with less clock skew to a large-area planar digital system. The second offers high data rate transmission, is reconfigurable and provides signal multiplexing capability in a LAN-based architecture.

Optical multiplexing is receiving less emphasis even though optical interconnects are receiving extensive development for rack-to-rack and even board-to-board communications, and optical interconnects have been proposed for ICs for almost 20 years [8.21]. We anticipate that the techniques described above for millimeter waves will be applied for optical interconnects as well, with optical technology showing appreciable technology advancements.

As described, various techniques are being explored to alleviate future interconnect bottlenecks when copper/low-κ interconnect technology scaling and additional interconnect levels on-chip become inadequate. One approach is to incorporate multiplexed interconnect technology with 3D wafer stacking to alleviate wiring constraints while minimizing the Si IC chip area. The packaging concept incorporates a novel active backplane, which contains millimeter wave transceivers, transmission lines and solder-bumped contacts to both a millimeter wave motherboard and to the stacked Si ICs [8.17].

### 8.2.3 Non-electrical Distribution of Signals and Clocks

Optical interconnect technologies have long been the on-chip interconnect technology of the future for replacing metal/dieletric global interconnects. A key advantage of optical interconnects is the extremely high information bandwidth that can be readily accommodated; a key disadvantage has been the chip area and power required for the conversion process, particularly the generation of the optical signal. However, as Cu/lowest-κ interconnects become a scaling limitation near 50nm MFS and progress continues in integratable optical components, on-chip optical interconnects become more viable [8.21, 8.22].

Optical interconnects used as microwave and millimeter wave mulitiplexed interconnects can be either radiated or guided. While radiated clocks may be feasible, multipath considerations require special packaging constraints. Guided wave alternatives are more likely for mainstream applications. Success has been demonstrated in optical wave guides at the silicon-$SiO_2$ level and also in polymer materials more appropriate as part of the interconnect hierarchy.

While a complete review of on-chip optical interconnects is beyond the scope of this chapter, a few key issues remain before acceptance by the IC processing and design community. These include:
1) size of integrated optical components and compatibility with standard silicon IC process flow
2) power consumption and optical link error rate-data rate performance
3) monolithic acceptance seems to require a fundamental technology breakthrough, either in Si-based optical sources or compound semiconductor growth on Si

For the near term, research options are being pursued aggressively. In the interim, progress will continue on chip-to-board optical technology for various applications, particularly were large information bandwidth is needed. While less well established, terahertz technology is another long range option. A key advantage of terahertz techniques is the inherent compatibility with time division multiplexing. The short pulse capability is promising, although traditional optical approaches appear more near term and are receiving more interconnect research funds.

# FUTURE DIRECTIONS IN IC INTERCONNECTS AND RELATED LOW-K ILD PLANARIZATION ISSUES

## 8.2.4 Non-Planar Integrated Assemblies – Three Dimensional (3D) Integration

Various 3D approaches have been demonstrated for innovative first-level packaging. One approach developed by Irvine Sensors Inc. is the "neo stack" technology [8.23-8.24], which uses bare dice that have been thinned and subsequently placed in a "neo wafer" with a special thermally conducting glue as shown in Figure 8.2. After polishing the neo-chips are diced and stacked onto a single mounting layer. The neo-chips are then interconnected with a network of conductors fabricated on the side of the neo-stack. Modules such as dense memory stacks, and optoelectronic components have been demonstrated [8.24-8.25]. Other 3D chip stacking approaches have also been developed by various companies, such as 3D Plus Electronics for light-weight heterogeneous cubes [8.26] and Dense-Pac Microsystems for memory devices [8.27].

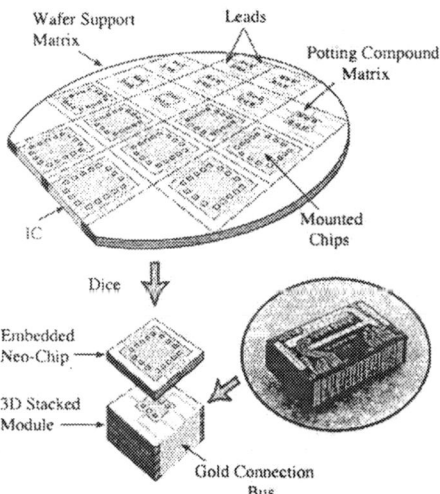

*Figure 8.2* The Irvine Sensor Neo-Stack process is based on a wafer-scale production of resin-embedded chips. After dicing, the chips are stacked and side-interconnected. Figure modified from [8.22], and used with permission.

Most methods of die/chip-stacking involve peripheral interconnection, which are potentially unreliable and expensive. These chip-stacks are not a complete 3D integration, but achieve folding of a large 2D chip into a 3D shape. The long wires on each chip are not reduced, as the side wires connecting the chips are still long. For example, an interconnect that travels to the perimeter from the interior of a chip is required and another interconnect needed to travel back to the interior of another chip on a different level. To shorten this (clock or signal) distance, a more direct path could be used if vias could be fabricated in the interior of the stack.

One approach, called "stacked wafer-level packaging" technology, has been proposed by Tru-Si Technologies, based on a through-silicon contact technology as shown in Figure 8.3 [8.25]. Contacts are embedded in the wafer from the front side, after wafer processing, the wafer is thinned to reveal the contacts through the back of the wafer. The stacking starts with an interposer wafer with through contacts that eventually will be the external I/Os. A device wafer is bonded face-to-face with the interposer, and the device wafer is thinned to expose through-silicon contacts, which act as bumps. Another device wafer is bonded face down onto those "bumps" and thinning and bonding can be repeated as needed. However, the thinned Si wafer is still quite thick (> 70 μm) and the size of the vertical interconnect is large (> 100 μm).

3D interconnect concepts have been proposed that further reduce the length of long wires, thus increasing the performance and functionality of advanced ICs. These include front-end processing using recrystallized semiconductors or wafer bonding using vias to internally (not just peripherally) connect the stacked wafers [8.29-8.32]. Wafer bonding using copper as a bonding agent [8.29-8.30] or using low-κ dielectric glues [8.31-8.32] appears feasible.

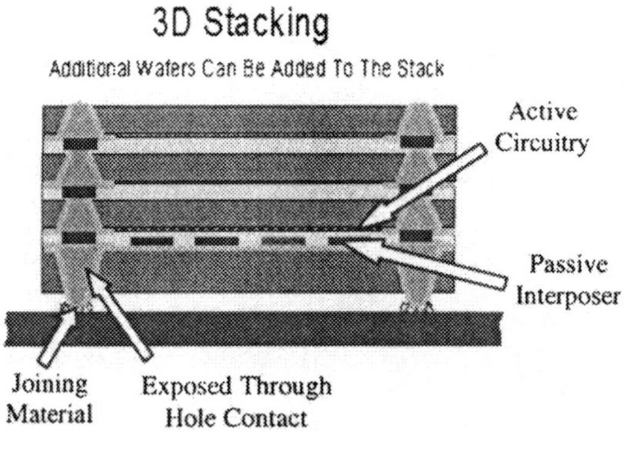

*Figure 8.3* Tru-Si 3D-stacked wafer-level packaging. Figure modified from [8.28].

Figure 8.4 depicts face-to-face and face-to-back wafer stacking cross-sections. The aligning, bonding, thinning and via-interconnecting processes required in the wafer bonding approach can be repeated as needed. This approach can dramatically reduce the length of long wires, improving digital clock rates while decreasing dynamic power dissipation and minimizing Si

IC chip area. Some of the interconnects serve as thermal pathways to conduct heat from the bottom layer to the top where the thinned wafer provides a short thermal path. In addition, high thermal dissipation cores (e.g. the processor) can be well heat sunk at the bottom (and/or top) of the stack.

The 3D wafer bonding technology with wafer-level vertical interconnects (high aspect ratio vias and damascene-patterned interconnects) has important implications in IC design. While conventional back-end-of-the-line (BEOL) technology needs to be extended [8.31, 8.32], an extension referred to as tail-end-of-the-line (TEOL) processing, design advantages and design tool needs can be projected. While these aspects depend upon application specific IC (ASIC), system-on-a-chip (SOC), mixed-signal, optoelectronic, microelectromechanical system (MEMS) and/or sensor applications, the issues can be delineated by considering two important areas -- intellectual property (IP) core-based design and high speed processor design [8.17].

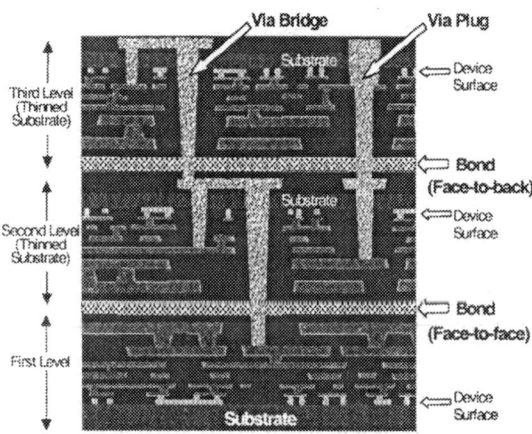

*Figure 8.4* 3D Integration using wafer bonding with dielectric glues [8.17, 8.32]. Figure modified from IBM Web page

Design reuse of IP cores (complex macrocells) is a growing response to design and verification crises, in which previously developed and performance proven designs are incorporated in new ICs. This concept extends the well-established use of in-house design libraries, but the extension to importing designs from IC foundries or design houses is an ongoing evolution for ICs. Licensing of fully verified IP cores can reduce time-to-market of new products significantly, as well as reducing new product risk. However, CAD tool compatibility, different foundry design rules and technology capabilities and industry structure issues will inhibit wide acceptance of full mix-and-match capability in the near future [8.17, 8.33-8.34].

A 3D wafer bonding TEOL capability would allow extension of the current approach to hard IP cores. Instead of licensing cores, full wafers could be purchased of embedded processors, digital signal processors (DSPs), memory (SRAM and DRAM), alternative non volatile memory (NVM), embedded wireless networks and other available building blocks. In one sense, the IP core approach is extended/replaced by wafer purchases, requiring both a common die size (anticipated to be a sub-multiple of a stepper field) and a common means of vertical interconnection (either peripheral or within-die interconnection requiring appropriate design rules).

The advantage of peripheral vertical interconnects is that the various subelements in the planar stacked system (PSS) design are unchanged except for the traditional bond pad areas; the disadvantage is that the PSS can become input/output (I/O) limited. The advantage of within-die interconnection is that interconnect-limited performance enhancements are achieved and I/O limitations are relaxed, at the expense of design rule constraints which could impact the advantage of "off-the-fab-line" purchase of wafers -- at least initially in such a technology use.

We envision that an early use of the 3D technology will be as a replacement for chip-on-chip packaging, which is becoming increasingly popular for memory (DRAM, SRAM or flash) attachment to a processor. Conventionally, the memory die is smaller than the processor die, and wire bonds are used for intrachip interconnection after die attach. The I/O limitation is controlled by the wire bond spacing and bond pad area (~75 um). With the 3D technology, peripheral interconnection between the memory and processor ICs can be done at the wafer level with a single level of damascene patterned interconnects; a conservative via interconnect spacing is 8 $\mu$m, allowing a significant interconnect density increase (additional layers could be necessary for signal and/or power distribution depending upon pad locations on the two ICs). Manufacturing yield considerations can be alleviated by using die map test results to stack wafers with matched good-die locations.

We hypothesize that for high performance digital systems and probably for mixed signal applications, the performance and I/O advantages of within-die interconnections will be important. For MEMS, smart sensors, biotech and other applications requiring heterogeneous technologies peripheral interconnects will probably be sufficient. For wireless applications, the situation is less clear, but within-die vertical interconnects are certainly desirable and could be mandatory in bandwidth-limited applications.

In the era of interconnect-limited processor design, appreciable effort has been expended to both understand the limitations of on-chip wiring and to properly simulate the effects on interconnect delay on synchronous digital processors [8.35-8.36]. All such work, at various levels of completeness and precision, conclude that the relatively small number of long-distance interconnects are the principal limiters to higher speed processors [8.37]. The overriding design rule is "keep the interconnects short" [8.38].

The 3D wafer bonding technology provides a means to replace long distance interconnects required in conventional 2D ICs with relatively short vertical interconnects. While the gain in interconnect delay is design specific, the reduction in long distance interconnection is clear. To obtain the most significant increase in performance, a high density of vertical interconnects is envisioned, requiring the development of 3D design rules.

## 8.3 3D WAFER-SCALE INTEGRATION USING DIELECTRIC BONDING GLUES AND INTER-WAFER INTERCONNECTION WITH COPPER DAMASCENE PATTERNING

A three-wafer stack depicted in Figure 8.4 illustrates a viable approach for 3D ICs pursued at Rensselaer as part of the Interconnect Focus Center. Initially two wafers are aligned and bonded face-to-face (or interconnect structure-to-interconnect structure). The top wafer is thinned to ~10 um and planarized, followed by an inter-wafer interconnection using copper damascene patterning (high-aspect-ratio (HAR) etching, via fill and CMP). Conventional BEOL processes are used, except for wafer-to-wafer alignment and bonding. An Electronics Visions Group (EVG) EV640 SmartView™ Aligner is used to provide 1-2μm alignment accuracy with 200 mm wafers, as has been demonstrated in our laboratory. This approach to monolithic, wafer-scale 3D integration is very BEOL process dependent and is unique in three aspects:

1) no handling wafer is needed to transfer a device layer after substrate thinning
2) no front-end processing is needed to apriori fabricate, in full or in part, the vertical inter-wafer vias
3) complete compatibility with conventional BEOL process is maintained (such as temperature <400°C and pressure <50psi).

Dielectrics with BEOL process compatibility, i.e., similar to ILDs, but with good bonding characteristics are an integral part of this approach to 3D integration. Wafer-scale planarization is also important to reduce the stresses introduced in the wafer bonding and thinning processes.

## 8.3.1 Wafer Bonding

While any new integration concept has technological challenges, a key challenge of our wafer-scale monolithic 3D integration is wafer bonding. Wafer bonding of a fully processed 200 mm wafer requires intimate contact over the entire wafer, and compatibility with both conventional BEOL processes and subsequent thinning and inter-chip interconnection. Currently available direct wafer bonding processes involve high temperature, high voltage and/or high pressure. Therefore, it is especially challenging to directly bond 200 mm wafers, on which ICs with several interconnect metal levels have been fabricated. The use of a polymeric "glue" layer, such as a low-κ dielectric, facilitates the wafer bonding for 3D integration.

The polymeric glue layer must provide a seamless interface and strong adhesion to prevent delamination, be sufficiently thin to minimize the via aspect-ratio, be sufficiently reactive to form a chemical bond at modest temperatures, and be thermally stable after bonding. In addition, the device wafer performance and reliability must not be affected by the processing. Five key requirements for the glue are: (1) good adhesion to the surfaces of both wafers and to glue layer itself; (2) no out-gassing after wafers are placed in contact: (3) high degree of cross-link network with high glass transition temperature (Tg); (4) low stress relaxation and creep to alleviate the mechanical stress; and (5) matched coefficient of thermal expansion (CTE) to wafers being bonded.

The candidates for bonding glue currently under evaluation at Rensselaer are spin-on amorphous polymers, such as FLARE™ (Poly aryl ether) [8.39], methylsilesequionexane (MSSQ) [8.40], benzocyclobutene (BCB) [8.41], and hydrogensilsesquioxane (HSQ) [8.42], as well as a vapor deposited polymer Parylene-N [8.43]. They can be deposited as thin films, have relatively well-known chemical and physical properties, and are cleanroom process compatible. In addition, these materials meet bonding requirements to some extent and are compatible with IC BOEL processes.

The bonding protocol largely consists of two main steps: preparation of the glue layers for wafer contact and wafer bonding. For spin on polymer materials (FLARE, MSSQ, BCB and HSQ), the glue layers are prepared as follows: 1) spin dispense polymeric material using a FlexiFab™ Coat Bake System; 2) pre-bake to remove solvents, volatile impurities and water formed by the condensation reaction, and to improve the degree of surface planarization. Parylene-N is deposited using a SCS multi-wafer deposition system (PDS 1050) with a di-$p$-xylyene dimer. Vapor deposition of Parylene-N is a solvent-free dry process without outgassing, and a very uniform film is obtained.

Wafer bonding is conducted using an EVG501 wafer bonder. Two wafers with glue layers are placed glue-to-glue, separated by "flags", and clamped together as illustrated in Figure 8.5. The wafer pair is loaded into the EVG501 bonding chamber, which is pumped down to approximately $10^{-4}$ mbar. The bonder chucks are heated to a given temperature while the mechanical spacers (flags) keep the wafers physically separated. The bond is originated by pressing in the middle of the top wafer to create an initial point of contact, using the "wafer-bow" process depicted in Figure 8.5(a). Upon removal of the flags (Figure 8.5(b)), a uniform bonding down pressure is applied over the wafer pair. The wafer pair is then further heated to complete the bonding process. Finally, the wafers are slowly cooled and unloaded. The maximum processing temperature for the wafer bonding is 400°C, compatible with both conventional BEOL processes and subsequent inter-chip interconnection and packaging processes.

In the 3D integration process, the processed wafers are first aligned in an EVG640 alilgner prior to loading in the bonder chamber. Therefore, the flags are important to separate the two wafers and prevent any gas trapping in the rough surface or wafer bow between the wafers before and during the bonding processes. The down force is applied to remove any gaps between the glue layers on the top and bottom wafers, and to facilitate the cross-link reaction. Curing (the cross-link network formation or the intermixing of polymer chains) is the core step of the wafer bonding, which provides the bond between the two wafers. A holding time is needed during the bonding process to form a complete cross-linked network. A slow cool down process is crucial to minimize debonding due to the stress generated by coefficient of thermal expansion (CTE) differences between polymer and wafer.

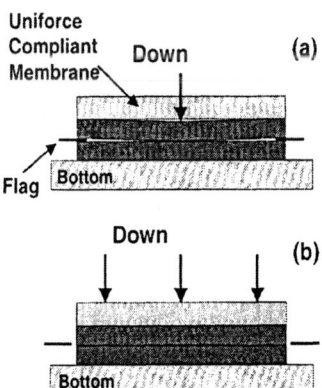

*Figure 8.5:* Schematic of bonding process (a) initial application of down force, (b) after flags removed.

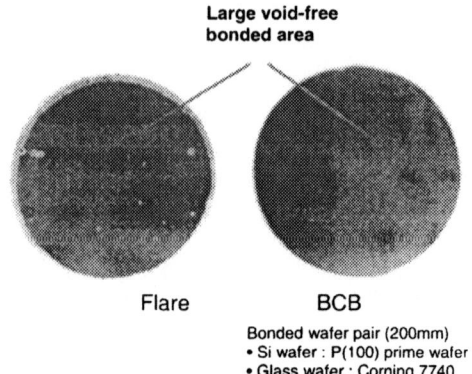

*Figure 8.6* Wafer bonding result using FLARE and BCB. Bonded wafer pair (200mm) : P (100) prime Si wafer and Corning 7740 glass wafer. Small voids are caused by particles and/or defects.

Wafer bonding results are particularly encouraging with FLARE and BCB. Figure 8.6 shows a photo of a pair of bonded wafers (200 mm Corning 7740 glass wafer to 200 mm p-type (100) prime silicon wafer) using FLARE and BCB. While the quality is quite good, some issues related to the bonding can be itemized as follows:
1) void generation due to outgassing of the remaining solvent or by-product created during the curing process
2) void generation due to defects or particles
3) wafer bow caused by mechanical stress

However, these bonded wafer pairs show little change in defect structure after thermal cycling to 400°C or after top-wafer thinning to 60μm as depicted in Figure 8.7. Results with BCB indicate no additional defects, with top-wafer thinning to 35μm achieved by back-grinding and polishing [8.44].

FLARE is a non-fluorinated poly aryl ether manufactured by Honeywell Inc. Without an appreciable carbonyl or water adsorption group, FLARE has very good resistance to water uptake, and no condensation reaction occurs during the process. The high degree of cross-linking during the curing process increases the structure rigidity and thermal stability. The cross-link network results in strong adhesion of FLARE to itself and to the films on surface of each wafer. In addition, weight loss is very small at curing temperature (less than 1%/hr). Thus FLARE is an attractive bonding glue which results in a larger fraction of bonded area and higher bond strength compared to alternative glues tested to date.

# FUTURE DIRECTIONS IN IC INTERCONNECTS AND RELATED LOW-K ILD PLANARIZATION ISSUES

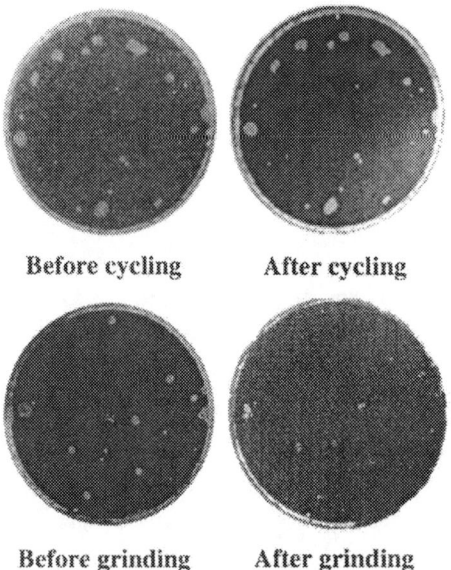

*Figure 8.7:* Initial bonding integrity results with FLARE (thermal cycling to 400°C at 50°C/min and back grinding to 60μm).

Benzocyclobutene (BCB) has been described in Chapters 2 and 4 and has also been used in packaging applications. While $T_g$ is less than desireable for some applications, the bonding properties are excellent.

### 8.3.2 Inter-wafer Interconnect

Subsequent to the wafer bonding, the top Si wafer is thinned and leveled. Four options have been explored to-date. All involve a first-step grind to 50-100 μm, followed by one or more of the following:
1) silicon wet etch
2) silicon spin etch
3) silicon atmospheric downstream plasma (ADP) etch
4) further grinding, followed by CMP

At this time, wafers have been thinned to 35 microns with good bonding integrity with each approach. Thinning to 10-15 microns has been demonstrated with both wet etching and spin etching. However, full characterization of silicon uniformity and built-in stress has not been reported.

The RIE requirements for the first step in the interconnection process include etching through silicon, silicon dioxide, silicon nitride, bonding glue

(BCB, FLARE, or other), and the ILD. Etching of patterned test structures indicates that such a process flow can be achieved in a dual-chamber system. The HAR vias required can be filled by both CVD and electroplated copper using suitable tantalum-based liners. However, process integration of these critical unit processes remains to be demonstrated with bonded wafer pairs [8.45].

The damascene patterning is completed with post-copper deposition CMP. The CMP capability has been demonstrated with an IPEC 372M rotary polisher, using commercial slurries, a Rodel IC 1400 k-groove pad and an OnTrak double-sided brush cleaner. Atomic force microscopy (AFM) images of patterned 200mm diameter wafers are shown in Figure 8.8 and optical micrographs of the first-generation via-chain test structures are shown in Figure 8.9 [8.46]. These results satisfy our requirements, although wafer-scale planarity remains to be evaluated.

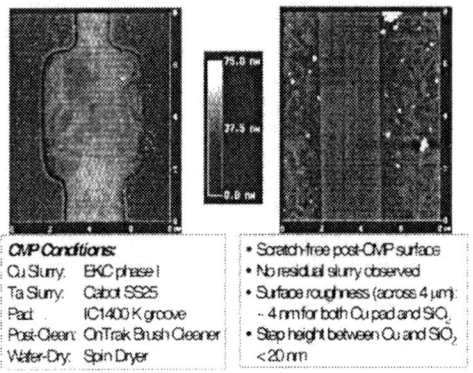

*Figure 8.8:* AFM Images of Damascene Patterned Wafers. From [8.45]

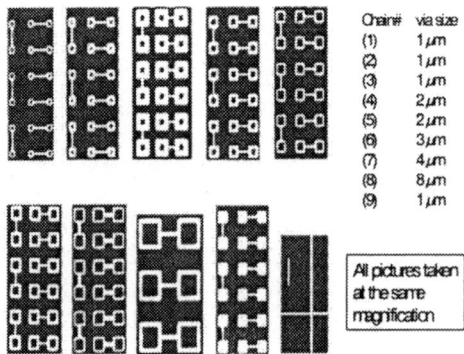

*Figure 8.9:* Microscopic Images of Damascene Patterned Wafers

### 8.3.3 Comparison with Other Wafer-Scale 3D Integration Technologies

Other approaches for wafer scale integration have been proposed and developed at Seimens (now Infineon), Lincoln Laboratories, Tohoku University and MIT [8.47-8.49, 8.29-8.30]. While a detailed comparison of the approaches and techniques is beyond the scope of this chapter, our approach has the following positive attributes:
1) the ability to accommodate wafer non-planarity and small particles
2) the ability to incorporate more than two wafers
3) the ability to incorporate micron-size inter-wafer interconnect
4) the lack of a required handling wafer

At this time the following concerns are being addressed:
1) HAR via etch requirement with different materials and the need to clean the bottom of the deep inter-wafer vias
2) bonding integrity degradation from subsequent processing such as wafer thinning and thermal cycling
3) wafer-scale planarity requirements to enable low stress bonding with thin glue layers while maintaining micron-scale alignment

While these results are promising, 3D wafer-scale integration is a relatively new technology. More research on alternative approaches is needed before the most attractive approaches can be fully delineated. Clearly the mechanical integrity of ultra low-$\kappa$ ILDs is one of many issues that must be addressed with 3D ICs based upon wafer bonding.

## 8.4 SUMMARY AND CONCLUSIONS

While there are many challenges to incorporate ultra low-$\kappa$ ILDs in IC interconnect structures, the industry is clearly concerned with "what's next?"; this chapter has presented the main alternatives to date, with emphasis on 3D ICs using wafer bonding as the most viable near-term alternative. Of course, predictions about the future in such a rapidly-evolving industry with such a technologically-creative work force are often incorrect. Perhaps the best use of this chapter is to help stimulate new approaches. Besides the technology-focused approaches presented, the impact of design architecture and applications is considered to be particularly important in driving the post copper/ultra low-$\kappa$ interconnect era, perhaps in directions that can't be delineated at this time.

## 8.5 REFERENCES

[8.1] S. Lopatin, A. Prewse, and R. Cheung, *Electrochem. Soc. Proc.*, **PV 99-31**, 221 (2000).
[8.2] D.P. Goetz, *Mater. Res. Soc. Proc.*, **566**, 51 (2000).
[8.3] A. Nickles, G. Leung, V. Mehan, H. Tam, P. McReynolds, G. Prabhu, and T. Osterheld, 18th International VLSI Multilevel Interconnect Conference (VMIC), 233 (2001).
[8.4] A. Simpson, L. Economikos, F.-F. Jamin, and A. Ticknor, *Mater. Res. Soc. Proc.*, **671**, M4.1 (2001).
[8.5] Y. Kamigata, Y. Kurata, K. Masuda, J. Amanokura, M. Yoshida and M. Hanazono, *Mater. Res. Soc. Proc.*, **671**, M1.3 (2001).
[8.6] S. Kondo, N. Sakuma, Y. Honna, Y. Goto, N. Ohashi, H. Yamaguchi, and N. Owada, IEEE International Interconnect Conference (IITC), 137 (2001).
[8.7] S.P. Mukherjee, J.A. Levert, and D.S. DeBear, *Matl. Res. Soc. Proc.*, **613**, E8.10 (2000).
[8.8] T. Hara, T. Taniguchi, K. Kinashita, H.Q. Li, J. Takamura, and T.C. Bristow, 18th International VLSI Multilevel Interconnect Conference (VMIC), 462 (2001).
[8.9] P.H. Yih, D.H. Wang, and S.H. Chiao, 18th International VISI Multilevel Interconnect Conference (VMIC), 59 (2001).
[8.10] R.J. Contolini, A.F. Bernhardt, and S.T. Mayer, *J. Electrochem. Soc.*, **141**, 2503 (1994).
[8.11] http://www.nutool.com
[8.12] D.M. Dhusari, M.D. Wedlake, P.A. Kohl, C. Case, F. P. Klemens, J. Miner, B.-C. Lee, R.J. Gutmann, J.J. Le, R. Shick, and L. Rhodes, *Mat. Res. Soc. Symp. Proc.*, **612**, D4.8 (2000).
[8.13] M. Lin, C.-Y. Chang, T.-Y. Huang, M.-L. Ling and H.-C. Lin, *Mat. Res. Soc. Proc.*, **612**, D4.7 (2000).
[8.14] M.G.M. Harris, P. Rich, and N. Rimmer, 18th International VLSI Multilevel Interconnect Conference (VMIC), 17 (2001).
[8.15] M. Hauder, et al., *Appl. Phys. Lett.*, **78**, 6,5 (2001).
[8.16] Private conversation with K. Saraswat, Stanford, 2002.
[8.17] R.J. Gutmann, J.-Q. Lu, R.P. Kraft, P.M. Belemjian, O. Erdogan, J. Barrett, and J.F. McDonald, "IP Core-Based Design, High-Speed Processor Design and Multiplexing LAN Architectures Enabled by 3D Wafer Bonding Technologies", DesignCon 2001: Wireless and Optical Broadband Design Conference, Feb. 2001.
[8.18] B.A. Floyd and K.K.O, IEEE Int. Interconnect Tech. Conf. (IITC), 248 (1999).
[8.19] B.A. Floyd, K. Kim, and K.K. O, ISSCC Dig. Tech. Papers, 328 (2000).
[8.20] M.F. Chang, V. Roychowdhury, L. Y, Zhang, Z. Wang, Y. Wu, P. Ma, C. Lin, and Z. Kang, IEEE Int. Interconnect Tech. Conf. (IITC), 21 (2000).
[8.21] D.A.B. Miller, *Proc. IEEE*, **88**, 728 (2000).
[8.22] R.H. Havemann and J.A. Hutchby, *Proc. IEEE*, **89(5)**, 586 (2001).
[8.23] The Interconnect chapter in *The International Technology Roadmap for Semiconductors* (ITRS): *1999 edition*, Semiconductor Industry Association, 1999.
[8.24] Keith D. Gann, "Neo-Stacking Technology", High-Density Interconnect, December 1999.
[8.25] V. Ozguz, P. Marchand, and Y. Liu, "3D Stacking and Optoelectronic Packaging for High Performance Systems", International Conference on High-Density Interconnect and Systems Packaging, April 2000.
[8.26] http://www.3d-plus.com
[8.27] http://www.dense-pac.com

[8.28] S. Savastiouk, O. Siniaguine, and E. Korczynski, "3D Wafer Level Packaging," 2000 International Conference on High-Density Interconnect and Systems Packaging, April 2000.

[8.29] A. Fan, A. Rahman, and R. Reif, *Electrochemical and Solid-State Letters*, **2 (10)**, 534 (1999).

[8.30] A. Rahman and R. Reif, IEEE International Interconnect Technology Conference (IITC), 157, (2001).

[8.31] J.F. McDonald, R. Kraft, J.-Q. Lu, T.-M. Lu, A. Kumar, T. Cale, P. Belemjian, O. Ergodan, A. Kaloyeros, and J. Castracane, 17th International VLSI Multilevel Interconnection Conference (VMIC), 90, (2000).

[8.32] J.-Q. Lu, A. Kumar, Y. Kwon, E.T. Eisenbraun, R.P. Kraft, J.F. McDonald, R.J. Gutmann, T.S. Cale. P. Belemjain, O. Erdogan, J. Castracane, A.E. Kaloyeros, "3-D Integration Using Wafer Bonding", in Advanced Metallization Conference (AMC), October 2000, to be published by MRS.

[8.33] J.-Q. Lu, Y. Kwon, R.P. Kraft, R.J. Gutmann, J.F. McDonald, and T.S. Cale, International Conference on Dielectrics and Conductors for ULSI Multilevel Interconnection (DUMIC), 235 (2001).

[8.34] J.-Q. Lu, Y. Kwon, R.P. Kraft, R.J. Gutmann, J.F. McDonald, and T.S. Cale, IEEE International Interconnect Technology Conference (IITC), 219, (2001).

[8.35] P. Zarkesh-Ha, and J.D. Meindl, IEEE Symp. on VLSI, 44 (1998).

[8.36] A. Rahman, A Fan, J. Chung, and R. Reif, Proceedings of International Interconnect Technology Conf. (IITC), 233 (1999).

[8.37] S.J. Souri, and K.C. Saraswat, Proceedings of International Interconnect Technology Conf. (IITC), 24 (1999).

[8.38] Comment J.D. Meindl, Georgia Institute of Technology.

[8.39] R.N. Vrtis, K.A. Heap, W.F. Burgoyne, and L.M. Roberson, *Mater. Res. Soc. Symp. Proc.*, **443**, 171 (1997).

[8.40] N.H. Hendricks, 17th International VLSI Multilevel Interconnect Conf. (VMIC), 17 (2000).

[8.41] S.F. Hahn, S.J. Martin and M.L. McKelvy, *Macromolecules*, **25**, 1539 (1992).

[8.42] D. Oben, P. Weigand, M.J. Shapiro and S.A. Cohen, *Mater. Res. Soc. Symp. Proc.*, **443**, 195 (1997).

[8.43] T.-M. Lu and J.A. Moore, *MRS Bulletin*, **22 (10)**, 28 (1997).

[8.44] R.J. Gutmann, J.Q. Lu and T.S. Cale, unpublished results.

[8.45] Conversations with Profs. Jim Castracane and Eric Eisenbraun Institute for Advanced Materials, University at Albany.

[8.46] Results obtained by G. Rajagopalan, M. Gupta and C.K. Hong, Rensselaer Polytechnic Institute.

[8.47] P. Ramm, D. Bonfert, H.Gieser, J. Haufe, F. Iberl, A. Klumpp, A. Kux, R. Wieland, Proceedings of the 2001 IEEE International Interconnect Technology Conference (IITC), 160 (2001).

[8.48] J. Burns, L. McIlrath, C. Keast, C. Lewis, A. Loomis, K. Warner, P. Wyatt, IEEE International Solid-State Circuits Conference, ISSCC, 268 (2001).

[8.49] K.W. Lee, T. Nakamura, T. One, Y. Yamada, T. Mizukusa, H. Hasimoto, K.T. Park, H. Kurino and M. Koyanagi, International Electron Devices Meeting (IEDM), 165 (2000).

# Appendix A

# EXPERIMENTAL PROCEDURES AND TECHNIQUES

This appendix describes the experimental techniques, procedures, and methods used in gathering data on the CMP of low dielectric constant thin films. The CMP operation including film preparation, slurry mixing, CMP parameters, and metrology techniques are described to supplement the CMP results presented in Chapters 4, 5, and 6.

## A.1 FILM DEPOSITION

The three low dielectric constant polymers used for CMP experiments, BCB-3022, BCB-5021, and SiLK were provided by Dow Chemical Company. The films were deposited as uniform coatings over the entire wafer. This technique, known as "blanket" wafer coating is used to investigate a specific property of a material. In microelectronics fabrication, blanket dielectric materials are deposited and then selectively removed in places to create interconnect trenches and vias.

BCB-3022 is an experimental grade of BCB that contains 99.5 % polymer material in solvent, and 0.5 % impurities. The BCB-3022 solution is a viscous liquid solution of precursor units dissolved in trimethyl benzene. The solids content of the solution is 32 wt%. BCB-5021 is a microelectronics application grade of BCB, that contains 99.9% polymer material and 0.01% impurities. The BCB-5021 also contains mesitylene as the precursor solvent, and has a solid content of 32 wt%.

An adhesion promoter was used prior to BCB coating, to increase the adhesion strength of the post-cure films [A.1]. Dow AP8000 was used in the experiments presented in Chapter 4. AP8000 is a molecule that contains silicon at one end of its chain to adhere well with the silicon wafer, and organic constituents at the other end of the molecule to adhere well with the organic polymer. The adhesion promoter was dripped on to the center of the wafer and spun at high speed, then baked for 1 minute at 100°C to drive off any remaining solvent. This step prevents evaporation of the AP8000 solvent during the BCB curing step, which can result in bubbles in the dielectric film.

After adhesion promoter coating and cure, two mL of either BCB-3022 or BCB-5021 solution were dripped on to the center of a 4 or 5 inch wafer positioned on a spin coating vacuum chuck. The wafer was then accelerated to the desired spin speed and held at that speed for a prescribed time. A spin speed versus thickness plot for BCB provided by Dow Chemical Company [A.2] is shown in Figure A.1. The BCB used for CMP experiments was spun at approximately 3500-4000 RPM, resulting in a coating 1000 – 1200 nm thick.

After spinning, the BCB was cured for 30 - 60 min at 250 - 300°C in a high vacuum; the base pressure of the vacuum furnace was approximately $10^{-8}$ torr. In the case of nitrogen curing, a constant flow rate of approximately 200 sccm of $N_2$ gas was cycled through the chamber to maintain an oxygen free ambient. The purpose of using high vacuum or nitrogen ambient in curing the films is to avoid oxidation of the films, which readily occurs at temperatures above 100°C in the presence of trace oxygen.

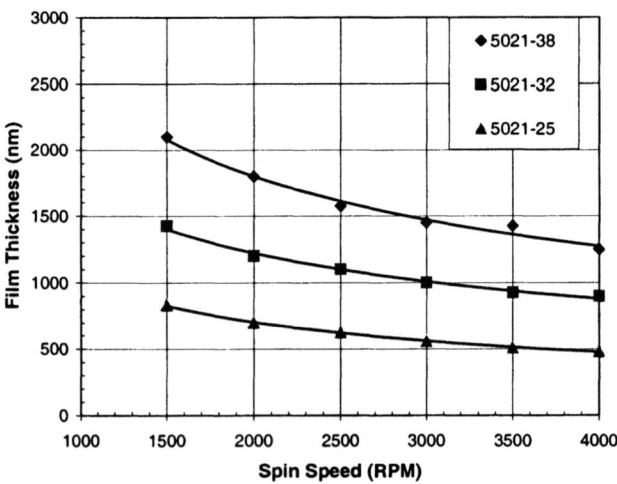

*Figure A.1.* BCB-5021 film thickness versus spin speed (from [A.2]).

The SiLK films were spin coated and cured by Dow Chemical Company. The wafers were cured in nitrogen at a temperature of either 400 or 450°C for 6 minutes. The SiLK was deposited without use of an adhesion promoter. After curing, the SiLK thickness is 1000 – 1100 nm. Table A.1 summarizes the different blanket polymer film coatings used for experiments.

Table A.1 – BCB and SiLK coatings used for experiments.

|  | Spin Speed (RPM) | Cure Time (min) | Cure Temperature (°C) | Final Film Thickness (nm) |
|---|---|---|---|---|
| BCB-3022 | 4000 | 30 | 250 | 1000 – 1200 |
| BCB-3022 | 4000 | 30 | 300 | 1000 – 1200 |
| BCB-5021 | 4000 | 120 | 300 | 1000 – 1200 |
| SiLK | 2700 | 30 | 400 | 1100 |
| SiLK | 2700 | 6 | 450 | 1100 |

Three OSG organosilicate glasses with varied carbon content and density have been investigated in this work. Film one is the best-known-method (BD1) version[*] of Applied Materials BlackDiamond[TM]. Films two and three are two versions[*] of Novellus Coral[TM*] denoted as C1 and C2. The OSG film deposition parameters and properties are listed in Table A.2, with BCB and SiLK listed for comparison.

The OSG films were deposited directly on 200mm silicon substrates by plasma-enhanced chemical vapor deposition (PECVD) using poly(methyl)-silane precursor gas and oxygen plasma. Deposition was performed at the film supplier's laboratory, on substrates from Texas Instruments. Concentrations of reactant gases and plasma power were adjusted to vary the carbon content in the deposited films. For example, BD1 was deposited at lower plasma power and lower wafer temperature, resulting in a lower relative carbon content and a slightly higher $\kappa$ of 2.85. C2 was deposited at higher plasma power and higher wafer temperature, allowing better reactant mobility and higher carbon content, also with $\kappa \sim 2.85$. C1 is deposited with higher plasma power than C2, causing a higher carbon content and a lower $\kappa$ of $\sim 2.70$.

---

[*] These experiments were done by one of the authors (CLB) during a summer internship at Texas Instruments; the process terminology used at TI is incorporated into the OSG film descriptions.
[TM] - BlackDiamond is a trademark of Applied Materials Inc., Santa Clara, CA.
[TM*] - Coral is a trademark of Novellus Systems, Inc., Santa Clara, CA.

Table A.2. Organosilicate film properties, with polymers listed for comparison.

|  | Deposition Method | Density (g/cm$^3$) | Carbon Content (approx. at%) | Dielectric Constant @1MHz |
|---|---|---|---|---|
| BD1 | PECVD | 1.43 [A.3] | ~ 20 | 2.85 |
| C2 | PECVD | 1.15 [A.3] | ~ 25 | 2.85 |
| C1 | PECVD | 1.19 [A.3] | ~ 28 | 2.70 |
| BCB | Spin-coat | 1.05 [A.4] | ~ 85 | 2.65 |
| SiLK | Spin-coat | 1.10 [A.4] | ~ 95 | 2.65 |

## A.2 SLURRY PREPARATION

Five slurries were used for BCB and SiLK CMP experiments. Each of the slurries was designed for the CMP of copper, to take advantage of the known material properties and chemical reactions of copper metal. Since alumina abrasives were exclusively used in copper CMP at the beginning of this research (and are still dominant today), alumina abrasive particles were used in the slurries. In accordance with the Pourbaix diagram for copper [A.5] and the isoelectric point of alumina [A.6], an acidic slurry pH between 1.0 and 4.0 was used. This pH range causes reduction of the copper surface and stable suspension of alumina abrasive.

The slurries are listed in Table A.3. All slurries consist of reagent grade nitric acid ($HNO_3$) and surface-active agents (surfactants) or Rodel QCTT1010 commercial slurry and reagent grade hydrogen peroxide ($H_2O_2$). The abrasive used for the control slurry and slurries 1-3 was degglomerate grade 0.05 μm $Al_2O_3$ from Meller Optics.

The control slurry is a standard copper slurry used to examine the role of surfactants during the CMP of hydrophobic polymer surfaces. When mixed with 0.1 – 0.3 vol% benzotriazole (BTA), the control slurry removes copper at a rate of 200 – 250 nm/min [A.7]. The acidic media dissolves the copper metal to form $Cu^{2+}$ wherever the copper is exposed to the slurry chemical. The BTA adsorbs on the copper surface to protect against etching. When the BTA is removed from the copper surface by the mechanical action of CMP, the slurry and abrasive remove copper metal.

Previous work suggests that surfactants enhance the contact of polymer materials and aqueous slurries, increasing CMP removal [A.8]. The contribution of surfactants is tested by comparison of the control slurry to slurries 1 and 2. Slurry 1 contains nonionic Triton-X 100 surfactant from Fisher Scientific, which has a hydrophobic octyl benzene group and a long polyethlyene-glycol hydrophilic chain, as shown in Figure A.2(a). This provides duality in the molecule without requiring charged states. The

Triton-X adsorbs on film and abrasive surfaces as the octyl benzene aligns itself away from water.

Table A.3. Slurries used for polymer CMP experiments.

|  | Main Components | Surfactant Additive | Oxidizer | Abrasive |
|---|---|---|---|---|
| Control | DI $H_2O$<br>$HNO_3$ (1 vol%) | None | None | $Al_2O_3$<br>0.05 μm |
| Slurry 1 | DI $H_2O$<br>$HNO_3$ (1 vol%) | Triton-X nonionic (1 vol%) | None | $Al_2O_3$<br>0.05 μm |
| Slurry 2 | DI $H_2O$<br>$HNO_3$ (1 vol%) | DowFAX 3BO anionic (1 vol%) | None | $Al_2O_3$<br>0.05 μm |
| Slurry 3 | DI $H_2O$<br>Rodel QCTT1010 | Unlisted Commercial additive | $H_2O_2$ (3.3 vol%) | $Al_2O_3$<br>0.05 μm |
| Slurry 4 | DI $H_2O$<br>Rodel QCTT1010 | Unlisted Commercial additive | $H_2O_2$ (3.3 vol%) | $Al_2O_3$<br>0.30 μm |

Slurry 2 contains anionic DowFAX 3BO surfactant from Dow Chemical Company, which has a long hydrophobic hydrocarbon chain and two negatively charged sulfate hydrophilic groups, as shown in Figure A.2(b). Surfactant duality is achieved by the molecule charge polarity. DowFAX 3BO is prepared in its acidic form, in the presence of acidic hydrogen ($H^+$), to eliminate sodium ($Na^+$) contaminants. 3BO surfactant adsorbs on all surfaces and interacts with surface charges on the polymers and abrasive.

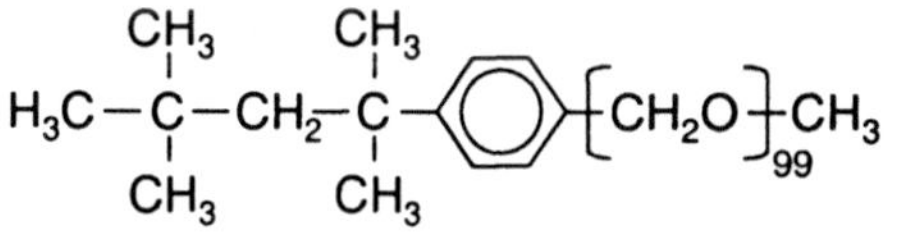

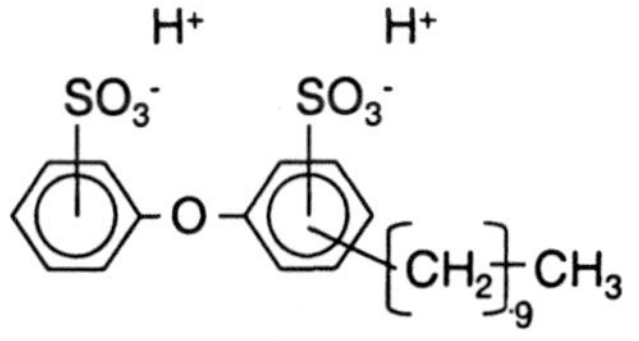

Figure A.2 Chemical structures of (a) Triton-X 100 and (b) DOWFAX3BO surfactants.

Slurries 3 and 4 contain the commercial QCTT1010 copper CMP chemistry manufactured by Rodel, Inc. These slurries were used to compare the oxidation / reduction properties of low-pH nitric acid slurries (slurries control, 1, 2, pH ~ 1.5) to slurries with milder pH and hydrogen peroxide oxidizer (slurries 3, 4, pH ~ 4.5). In addition, the impact of abrasive size on polymer CMP was examined by decanting the as-received slurry chemical and adding 0.05 μm $Al_2O_3$. Thus, slurry 3 contains the QCTT1010 chemical plus 0.05 μm $Al_2O_3$ from Meller Optics. TEM images of the Meller Optics abrasive and the Rodel abrasive (see Figures A.3(a) and A.3(b)) show slight differences in the abrasive texture and shape. In other work, the effect of the type of abrasive particles on copper CMP removal rate is discussed [A.8]. It is reasonable to assume that particle morphology may have a similar effect on polymer CMP, but in this study, the particles are assumed to differ only in size. The QCTT1010 slurries were diluted with deionized water to achieve an $Al_2O_3$ concentration of 1 wt%. All slurries were under constant agitation for 30 minutes prior to CMP to provide good abrasive suspension.

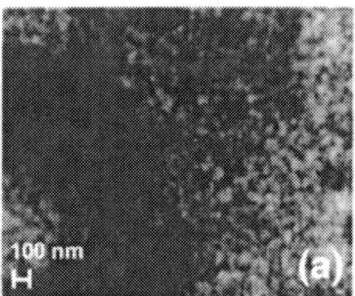 

*Figure A.3* 160,000x magnification TEM micrographs of (a) Meller Optics 0.05 μm $Al_2O_3$ and (b) Rodel QCTT1010 0.30 μm $Al_2O_3$.

The slurries used to polish OSG wafers are listed in Table A.4. Slurry A is a potassium hydroxide (KOH) based commercial silicon dioxide polishing slurry (Cabot SS11) diluted with deionized (DI) water to attain a solids content of approximately 3.5 wt% and a pH of 10.8. Slurries B, C, and D were mixed by gradually adding acid to Slurry A to attain pH values of 9.5, 7.7, and 6.0. Both propionic and citric acids have been used to adjust the slurry pH. The amounts of acid required to titrate the pH to the desired level were not measured, but were relatively small -- on the order of 10 ml (citric) to 50 ml (propionic) per liter of slurry. At a pH of 6.0, the silica slurries were sufficiently stable for immediate use, but the $SiO_2$ abrasive falls from colloidal suspension in less than two hours.

Slurries A-D were chosen for the blanket CMP of OSG to determine whether the mechanism for polishing organosilicates is similar to the mechanism for polishing $SiO_2$. Slurry E is a commercial copper polishing slurry consisting of alumina abrasive in acidic media. Slurry E is very similar to Slurry 4, the commercial QCTT1010 chemistry with $Al_2O_3$ abrasive. This slurry was chosen to examine the combined effects of acidic slurry environment and harder abrasive particles on OSG removal, to represent the possibility of some OSG material becoming exposed during the copper removal. All of the slurries were prepared using commercial mixtures plus the addition of DI water or dilute acids. No wetting agents or surfactants were separately added to the slurries.

*Table A.4* Slurries used for direct CMP of OSG and/or SiLK. The pH of the silica slurries is adjusted by the addition of organic acids.

|  | Abrasive | Abr wt% | pH | Material Designed to Polish |
|---|---|---|---|---|
| Slurry A | Colloidal silica | 3.5 | 10.8 | $SiO_2$ |
| Slurry B | Colloidal silica | 3.5 | 9.5 | $SiO_2$ |
| Slurry C | Colloidal silica | 3.5 | 7.7 | $SiO_2$ |
| Slurry D | Colloidal silica | 3.5 | 6.0 | $SiO_2$ |
| Slurry E | Alumina | 5 | 4.0 | Cu |

## A.3 CMP OF BLANKET FILMS

The polymer and OSG films were polished to measure removal rates in the slurries described. CMP tool parameters were selected to minimize the mechanical forces on the softer materials. Post-CMP metrologies were chosen to gain insight into the individual phenomena that impact the CMP performance of low-κ films.

### A.3.1 Polisher Setup

CMP of BCB and SiLK was performed on an IPEC/Westech 372M polisher with a single polishing head, using the primary polishing platen. Two pads were used for experiments, SUBA IV and IC 1400 k-grv, both manufactured by Rodel Inc. The SUBA IV pad is a softer pad with more fibrous structure, and the IC 1400 pad is a harder pad with a more dense structure. The "k-grooves" on the IC 1400 pad are shaped as illustrated in Figure A.4, to provide slurry chemical and abrasive transport through the pad to the wafer surface.

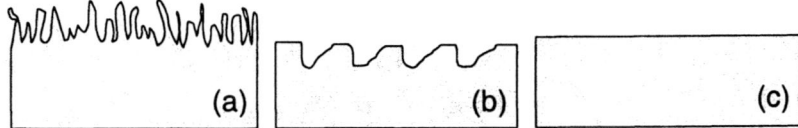

*Figure A.4.* (a) SUBA IV pad with fibrous structure, (b) IC 1400 k-grv pad with harder structure and grooves, (c) IC 1000 pad that is harder polyurethane material, with no grooves.

Wafer/pad relative velocity of 0.55 m/s (30 rpm carrier, 30 rpm platen) and downforce of 17.24 kPa (2.5 psi) were used for polishing polymer samples. The low pressure and speed were chosen in order to minimize the physical damage to the relatively soft polymers. Slurry was dispensed to the center of the polishing pad at a rate of 200 ml/min. In-situ pad conditioning was performed during polishing using a diamond honeycomb disk at a pressure of 0.03 kPa (0.2 psi). *In-situ* conditioning was used with the polymer samples since more consistent pad properties are maintained throughout multiple experiments when such conditioning is used [A.9].

CMP of blanket OSG films was performed using an Applied Materials Mirra CMP system. Wafers were loaded and polished using one of the slurries listed in Table A.4 for 30 seconds. The carrier membrane pressure (downforce) was varied between 14 and 21 kPa (2 – 3 psi), while the wafer/pad relative velocity was varied between 1.2 and 1.4 m/s (90 – 100 rpm platen and carrier). A Politex embossed pad from Rodel was used with slurries A-D, and an IC1000 pad (see Figure 3.4) from Rodel was used with slurry E. An *in-situ* pad conditioning step was used when polishing with slurry E and the IC1000 pad, but no pad conditioning was used with the soft Politex pad. Following CMP, the 8 inch wafers also cleaned using a DI-water brush-scrub on an OnTrak/Lam double-side brush cleaner.

### A.3.2 Timed Polishing

A set of polishing procedures is used for each series of low-κ CMP experiments. After a new pad is placed on the tool, new pads are "broken in" for 10 minutes using a silicon wafer and water. The pressure of CMP helps to penetrate the pad with water and establish an elevated temperature environment closer to actual CMP operation. Next a silicon wafer is polished for 1 minute on the tool while the slurry of choice is dispensed to the center of the pad. This ensures that the pad becomes saturated with the slurry, so that the polymer wafers may be introduced to the pad without any lead time for slurry equilibrium.

Next the sample is polished using the slurry for a designated amount of time, with the pad conditioner operating during the timed polish (*in-situ*). Timed polishes are used to calculate material removal rates. Following each timed polish, the wafer is brushcleaned with deionized water and spun dry, and the polymer film thickness is measured. The brushcleaning operation is vital for removal of abrasive particles from the polymer surface. Following cleaning and measurement the wafer returns to the CMP tool for another timed polish / clean / measurement cycle.

## A.4 POST-CMP METROLOGY AND FILM CHARACTERIZATION

Measurements made following CMP provide the fundamental information that is used to understand the fundamentals of low-κ CMP. Due to the small thickness (~ 1 μm) of the thin polymer films, advanced techniques must be used to extract film information. Accurate direct physical measurements are difficult at such small dimensions. Light scattering techniques are used to determine film thickness, atomic force measurements are used to obtain a picture of the polymer surface topography, and X-ray scattering from atoms is used to determine the chemical composition and bonding structure of the top few atomic layers of the polymer surface.

### A.4.1 Thickness Measurements

Film thickness is measured to determine film removal during the CMP process, and to characterize the uniformity of film removal. Polymer thicknesses were measured prior to CMP using a null ellipsometer from Rudolph Research and an interferometer from Nanospec. Seventeen points were measured on each wafer, according to the measurement template shown in Figure A.5(a). With basic knowledge of the film thickness from the correlation in Figure A.1, ellipsometry can be used to calculate an accurate refractive index for a translucent film. Refractive index values of 1.56 for BCB and 1.63 for SiLK are obtained from ellipsometry in accordance with the pre-CMP film thickness. These values are then used for multiple-wavelength interferometry measurements. The tool measures the interference pattern of light varied over a wavelength range of 400 – 800 nm, and uses the refractive index of the material to calculate its thickness.

OSG film thickness and refractive index (RI) were measured using a Tencor F5 measurement tool in spectroscopic ellipsometry mode at 21 measurement sites per wafer, according to the measurement pattern shown in Figure A.5(b). Spectroscopic ellipsometry is an ellipsometry scan that is performed over multiple wavelengths of light, from 400 – 800 nm. This provides a more complete solution for film thickness and refractive index, as information from multiple wavelengths are compared and combined.

Each optically measured thickness was then physically confirmed by a physical measurement using profilometry. A diamond scribe was used to scrape away a portion of the film to form a narrow trench. The depth of this trench is then measured using a microscopic tip that is dragged across a section of the trench, showing the depth (and corresponding thickness) of the polymer film.

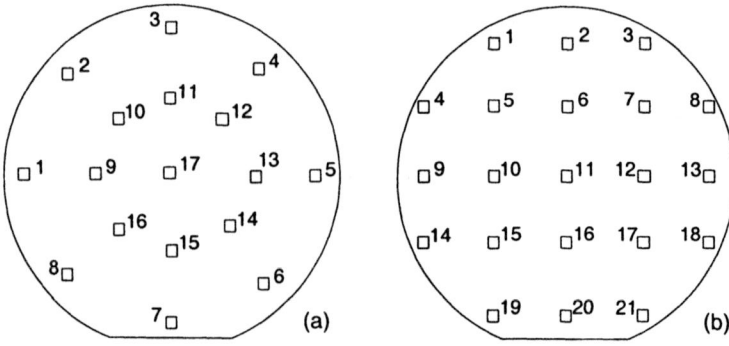

*Figure A.5.* Measurement templates used for (a) polymer thickness interferometry and (b) OSG thickness ellipsometry before and after CMP.

After verifying the thickness by multiple techniques, the Nanospec Interferometry technique was used exclusively due to its relative quickness and ease of wafer handling. After initial measurement, the films are polished, cleaned, and re-measured. Subtraction of pre- and post- CMP thicknesses is used to calculate polymer CMP removal rates.

### A.4.2 Atomic Force Microscopy

Following CMP, the films were measured by atomic force microscopy (AFM). AFM uses a very small tip geometry and piezoelectric motion to raster across a sample surface, measuring the forces imparted on the AFM tip by the individual atoms and molecules at the sample surface. The apparatus consists of a tiny pyramidal tip at the end of a cantilever. A laser shines on the back of the cantilever, measuring its deflection as the tip is traced across a surface, as shown in Figure A.6 [A.10]. The most minute deflections of the cantilever, due to atomic forces or small-scale topography, are measured and translated into a three-dimensional picture of a surface. A sample AFM image of a surface is shown in Figure A.7.

Using AFM, the surface characteristics of dielectric films before and after CMP can be measured on the scale of nanometers, or atomic clusters. The small circles in Figure A.7 are molecular clusters of material. This high resolution technique has no difficulty measuring nanometer and angstrom dimension surface topography. The research AFM used was an AutoProbe CP from Park Scientific Instruments, equipped with a type A microlever. The microlever was rastered in contact mode at a scan rate of 1 Hz.

# EXPERIMENTAL PROCEDURES AND TECHNIQUES

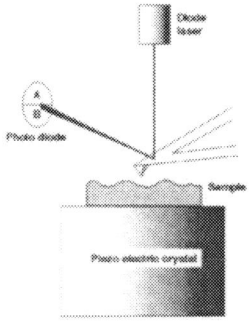

*Figure A.6.* A schematic diagram showing the basic principles of AFM operation (from [A.10]).

*Figure A.7* – A sample AFM Scan of a hard disk surface. The scan shows molecular-scale topography, with light areas being protrusions and dark areas depressions (from [A.10]).

Initial scans of 5 μm x 5 μm were measured in order to ensure tip contact and absence of abrasive particles on the sample surface. Tip contact is tested due to the nature of the polymer surface, which is easily charged by static energy, which makes the AFM tip lose contact. It is also important following CMP to make sure that abrasive particles do not remain on the sample surface since the dimensions of an $Al_2O_3$ particle are much larger than the dimensions of the polymer surface, causing a large discontinuity in the polymer surface scan.

Once a 5 μm x 5 μm area was successfully measured, three 1.5 x 1.5 μm scans were recorded within that 5 μm x 5 μm area. These smaller scans were analyzed for root mean squared (RMS) roughness and scratch depth. RMS roughness is defined by Eq. A.1 as the standard deviation in the height measurements recorded in the full scan area [A11].

$$R_{rms} = \sqrt{\frac{\sum_{n=1}^{N}(z_n - \bar{z})^2}{N-1}}, \text{ where } \bar{z} = \text{mean z height} \tag{A.1}$$

## A.4.3 Nanoindentation

The mechanical properties of the BCB and SiLK films were measured before and after CMP using a MicroMaterials nanoindenter, model 550. The tool provides a constant load to the film surface using a 4-sided pyramidal tip with a radius of approximately 0.25 μm. A sample tip geometry showing pyramidal form and basic tip dimensions is shown in Figure A.8. The load results in a compression of the polymer layer, and a measurement of the tip depth into the material. This information may be used to calculate the thin film hardness and elastic modulus.

The nanoindentation conditions used were 0.05 mN initial load, 0.07 mN/s load rate, 10 s dwell time at each load step, and 25 μm spacing between sequential test sites. Five sites per sample were measured to generate load/depth information. Maximum probing depth was set to 250 or 400 nm. The resulting load vs. depth curves are used to calculate the mean hardness and elastic modulus of the thin films, to see how CMP in different slurries impacts them mechanically [A.12].

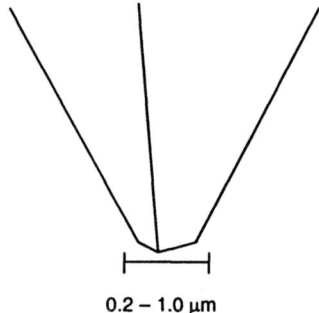

*Figure A.8.* Schematic enlargement nanoindentation measurement tip.

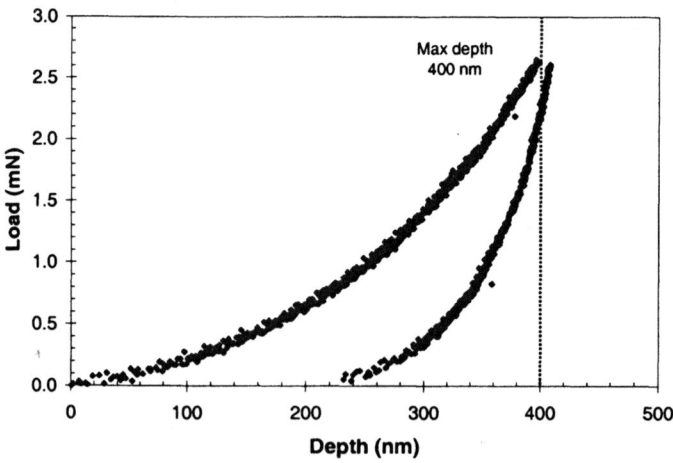

*Figure A.9* Example nanoindentation of load vs. depth data (from [A.13])

### A.4.4 X-ray Photoelectron Spectroscopy

Selected wafer samples also were measured following CMP by surface and depth profiling X-ray photoelectron spectroscopy (XPS). The XPS technique measures the energies of photoelectrons emitted by surface atoms when they are bombarded by X-rays from a monochromatic source, as illustrated in Figure A.10. The energies of the emissions are characteristic of the binding energies of the surface atoms, which can be correlated to the atomic number and chemical bonding of the atoms at a surface [A.13].

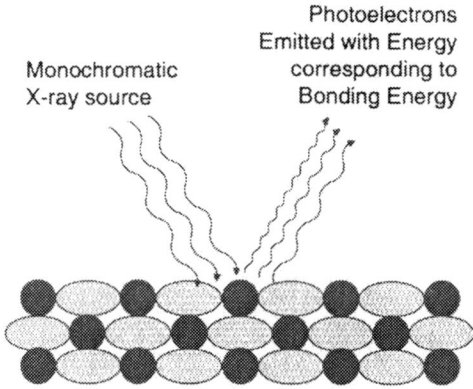

*Figure A.10.* Schematic diagram depicting XPS measurement of film surface atoms.

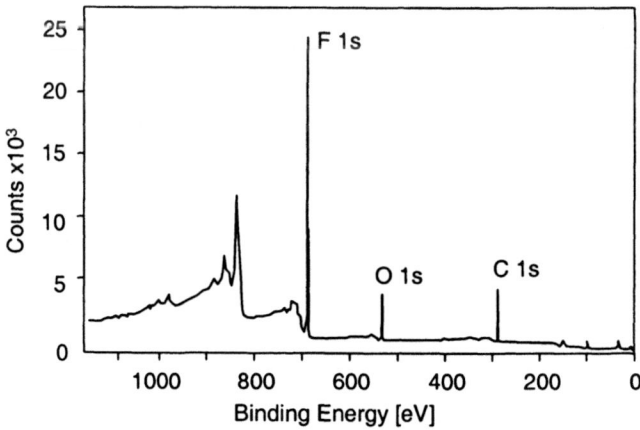

*Figure A.11.* Sample XPS spectrum for poly(tetrafluoro)ethylene (PTFE), showing binding energy peak locations according to detected photoelectron energy (from [A.14])

A sample XPS spectrum for the fluorinated polymer poly(tetrafluoro)ethylene (PTFE) is provided in Figure A.11. The spectrum illustrates the basic characteristics that are observed from an XPS measurement. The x-axis shows the binding energy of the surface atoms, as measured by the emitted photoelectrons. The peaks appear at different energies that are characteristic of different atomic bonds. This spectrum shows the presence of the main bonding constituents: a carbon peak corresponding to the 1s orbital bond energy, an oxygen 1s peak, and a fluorine 1s peak. A survey spectrum such as this is useful for comparing the relative amounts of a bound atomic species at the sample surface.

### A.4.4.1 Angle-Resolved Surface Analysis

Surface composition and bonding information for low-κ films before and after CMP was measured using angle-resolved XPS. Typical XPS measurements use a single angle and X-ray power, to gather information from the first 5 – 10 nm of the film surface. When the angle of the incoming X-rays is varied from 90 degrees to a glancing geometry, a shallow surface profile may be attained for bond energy and chemical composition. The geometry is shown schematically in Figure A.12.

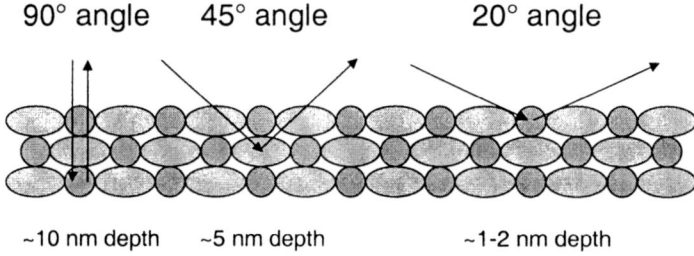

*Figure A.12* – Schematic representation of angle-resolved XPS near-surface depth profiling.

The incident angles used were 20, 45, and 90 degrees, using an RBD Enterprises SAGE 100 photoelectron spectroscopy tool with an Al $K_\alpha$ X-ray source at 350 W and 20 mA. The angles used result in near-surface polymer and OSG depths of approximately 1-2, 5, and 10 nm. The monochromatic X-ray source had a sample area of 1300 μm x 500 μm.

*A.4.4.2 Sputtered Depth Profiling*

Additional XPS results were taken to determine depth profiles of chemical composition and bonding in the polymers following CMP. The samples were measured using fixed-angle surface XPS in a vacuum chamber equipped with an ion gun for sputtering material. After measuring data from the polymer surface, a low energy ion sputter is used to remove a layer of the material and again measure the composition and chemical bonding.

The sputter gas used was argon, accelerated towards the polymer surface at a power of 300 W at 4 kV. XPS data was taken using a Mg $K_\alpha$ anode as the monochromatic x-ray source. Sputtering is measured in timed cycles, not in measured depth, due to the different sputter rates of BCB and SiLK materials. Twenty-five sputter cycles were used for each depth profile, resulting in a final depth into the polymer interior of approximately 700 nm for BCB and 1100 nm for SiLK polymer. The BCB sputter rate is lower due to the presence of silicon in the polymer chain. After each sputter cycle, the surface composition and bonding character were measured by surface XPS.

### A.4.5 Transmission FTIR

Bulk film chemistry of OSG films before and after CMP was measured using Fourier-transform infrared spectroscopy (FTIR). In infrared spectroscopy, IR radiation is passed through a sample. Some of the infrared radiation is absorbed by the sample and some of it is passed through (transmitted). The resulting signal has information about every infrared frequency which comes from the source. This means that as signal is measured, all frequencies are being measured simultaneously. Individual absorption at a single wavelength of IR energy is calculated by taking the fast Fourier Transform of the data signal. The resulting spectrum represents the molecular absorption and transmission of the sample, providing information about molecular structure and bonding throughout the bulk of a thin film sample [A.18].

FTIR measurements were performed over a wavelength range of 400 - 4000 $cm^{-1}$ before and after CMP using a Nicolet ECO1000s FTIR Semiconductor Wafer Analyzer. A sample spectrum for silicon dioxide dielectric is provided in Figure A.13. The figure shows the x-axis of IR wavelength, and the y-axis of measured absorbance. Thus peak location identifies molecular structures by bond energies, and peak height reflects the relative quantity of that molecular bond in the film. Thin film measurements are challenging, since the low-κ film comprises the top ~1 micron on a wafer which is ~1000 microns thick. For this reason, prior to film deposition, a background spectrum for a bare silicon wafer is measured, then subtracted from the measurement of the wafer and low-κ film.

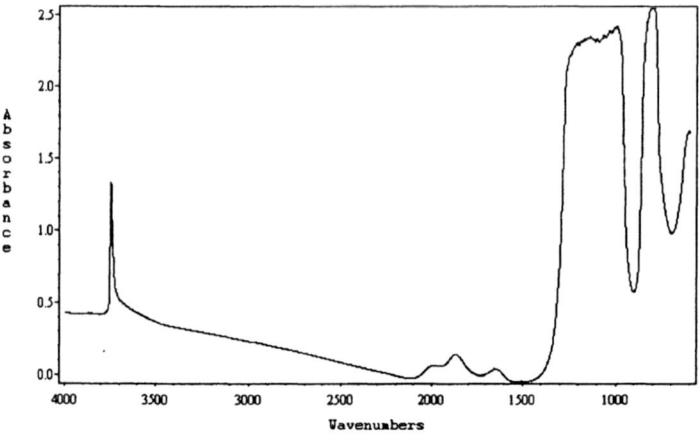

*Figure A.13.* A sample FTIR spectrum for silicon dioxide (from [A.16])

## A.5 REFERENCES

[A.1]   D. Pantelidis, H.-J. Lee; J. C. Bravman, in: S.A. Ringel, E.A. Fitzgerald, I. Adesida, D. C. Houghton (Eds.), *III-V and IV-IV Materials and Processing Challenges for Highly Integrated Microelectronics and Optoelectronics*, Boston, USA, November 30 – December 3, 1998, Materials Research Society Symposium Proceedings 535, 165 (1999).

[A.2]   Edward Shaffer, Dow Chemical Company, Facsimile Communication, 05/04/98.

[A.3]   C. Jin, SEMATECH, Austin, TX, private communication.

[A.4]   E. Shaffer, Dow Chemical Company Microelectronics Applications, Midland, MI, private communication.

[A.5]   M. Pourbaix, Atlas of Eletrochemical Equilibria in Aqueous Solutions, Oxford: Pergammon Press (1975).

[A.6]   B. Jirgensons, M. E. Straumanis, A Short Textbook of Colloid Chemistry, New York: MacMillan (1962).

[A.7]   M. T. Wang, M. S. Tsai, C. Liu, W. T. Tseng, T. C. Chang, L. J. Chen, M. C. Chen, *Thin Solid Films*, **308-309**, 518 (1997).

[A.8]   B. C. Lee, Model Slurries for Chemical Mechanical Planarization of Copper, Ph.D. Thesis, Rensselaer Polytechnic Institute, Troy, NY (2000).

[A.9]   K. Achuthan, J. Curry, M. Lacy, D. Campbell, S. V. Babu, *J. Elect. Mat.*, **25(10)**, 1628 (1996).

[A.10]  I. Larson, C. J. Drummond, D. Y.C. Chan, F. Grieser, *J. Am. Chem. Soc.* **115(25)**, 11885 (1993).

[A.11]  Park Scientific Research, AutoProbe CP Users Manual Chapter 9 – Understanding Surface Measurement Parameters, 9-53.

[A.12]  C.-W. Liu, B.-T. Dai, C.-F. Yeh, *J. Electrochem Soc.*, **142(9)**, 3098 (1995).

[A.13] L. C. Feldman, <u>Fundamentals of Surface and Thin Film Analysis</u>, New York: North-Holland (1986).
[A.14] S. Qu, T. Rosenmayer, P. Xu, P. Spevack, *Mater. Res. Soc. Symp. Proc.* 581, 375 (2000).
[A.15] Nicolet Instrument Corporation, <u>Introduction to Fourier Transform Infrared Spectrometry</u>, Madison, WI: Nicolet Instruments (2000).
[A.16] D. L. Sullivan, J. G. Ekerdt, *J. Catalysis,* **172**, 64 (1997).

# Appendix B

# XPS DEPTH-PROFILE DATA

The following figures show X-ray photoelectron spectroscopy (XPS) depth-profiling results for BCB and SiLK polymers before and after CMP. The purpose of the measurements was to determine the depth of penetration (if any) of slurry chemistry into the polymer films during CMP. To observe this, films were measured prior to CMP (BCB Figure B-1 and SiLK Figure B-3) as unpolished references.

Two films were then polished for several minutes in slurry 1, a nitric acid and surfactant slurry with 1 wt% 0.05 micron alumina abrasive. The results (BCB Figure B-2 and SiLK Figure B-4) show that the nitric acid slurry causes chemical change to the polymers only within the first depth-sputtering cycle, and the chemical content is constant as sputtering continues. This means that any oxidation or chemical attack by the nitric acid is limited to the top 50 – 100 angstroms of the films, and does not penetrate into the bulk of either BCB or SiLK.

Two additional films were polished for several minutes in slurry 4, the commercial QCTT1010 slurry with 1 wt% 0.30 micron alumina abrasive. The results (BCB Figure B-3 and SiLK Figure B-6) show that the QCTT1010 chemistry also causes chemical change to the polymers that is limited to the first depth-sputtering measurement cycle. Any oxidation or chemical attack by the QCTT1010 slurry, which allows 6 – 30x higher removal rate than the nitric acid slurry, is also limited to the top 50 – 100 angstroms of the films, and does not penetrate into the bulk of either BCB or SiLK.

This data supports the hypothesis of synergistic chemical-mechanical CMP removal of both BCB and SiLK polymers by an altered-surface layer reaction that is limited to the surface region of the films.

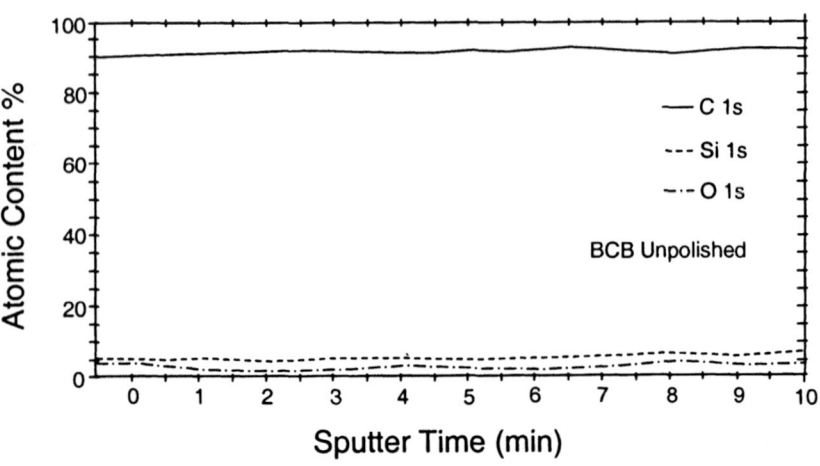

*Figure B.1* Depth profiling XPS spectrum for BCB before CMP

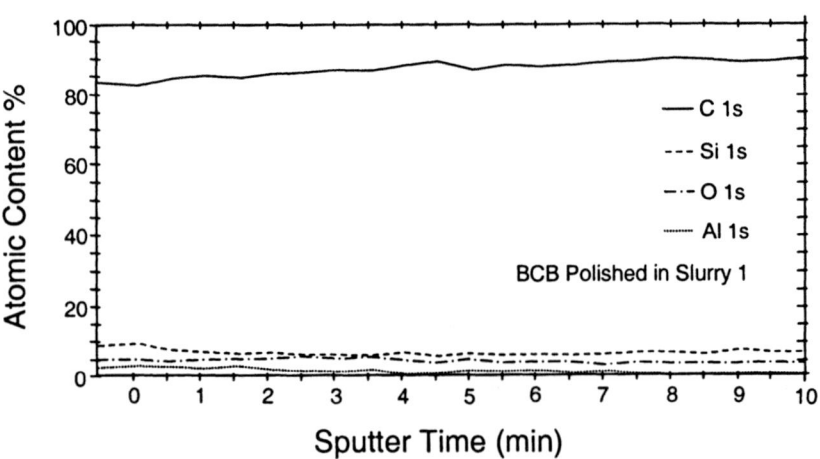

*Figure B.2* Depth profiling XPS spectrum for BCB after CMP with slurry 1

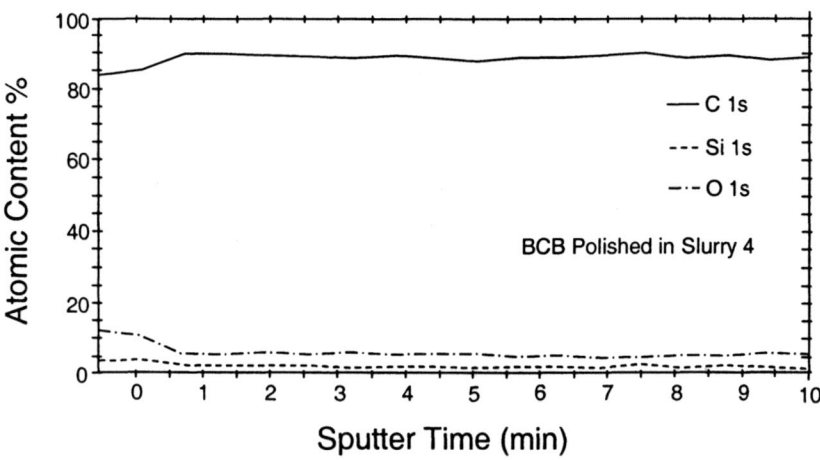

*Figure B.3* Depth profiling XPS spectrum for BCB after CMP with slurry 4

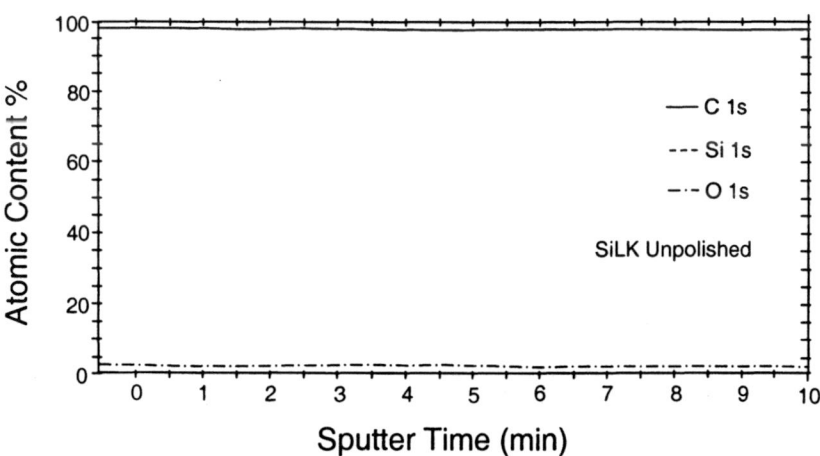

*Figure B.4* Depth profiling XPS spectrum for SiLK before CMP

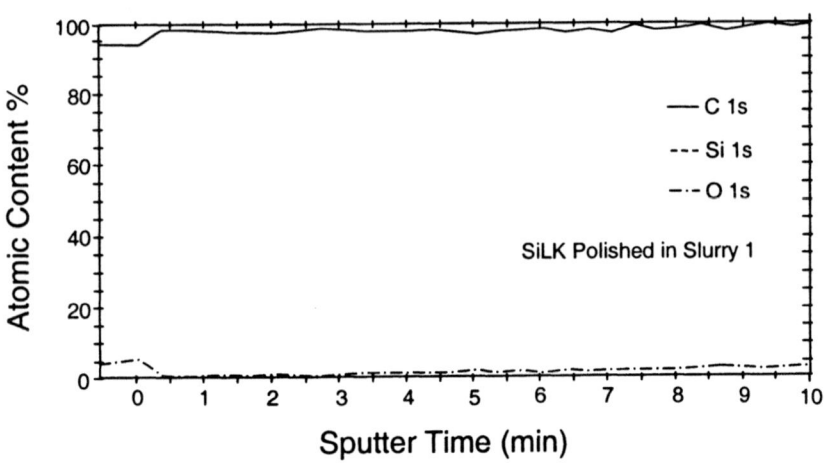

*Figure B.5* Depth profiling XPS spectrum for SiLK after CMP with slurry 1

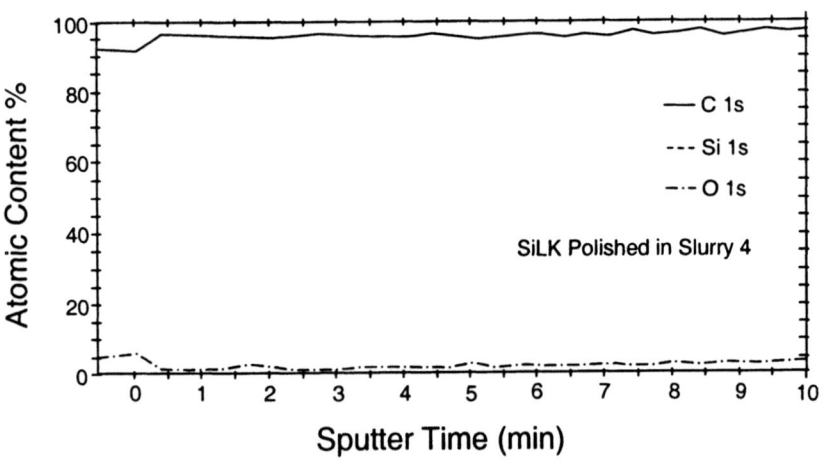

*Figure B.6* Depth profiling XPS spectrum for SiLK after CMP with slurry 4

# Appendix C

# CMP DATA FOR ANOMALOUS SiLK REMOVAL BEHAVIOR

The CMP data discussed in Chapter 4 show SiLK CMP removal rate measured as a function of CMP slurry chemical concentration, pressure, and velocity. Each data point is the result of an average thickness measurement from an intact post-CMP polymer film. The soft polymer films survive the CMP process due to the synergistic chemical-mechanical CMP process described in Section 6.4.1. When process variables are changed so that the balance between chemistry and mechanical removal is shifted, either low removal rate or high film damage can occur. Table C.1 lists several experimental CMP conditions in which the synergism of chemistry and mechanical removal is shifted and undesirable results are obtained.

*Table C.1* Data for $C_{active|inlet}$, $P$, $V$, or $d_p$ which result in an imbalance in the synergy between chemical alteration and physical film removal

| $P$ (psi) | $V$ (rpm) | $d_p$ (nm) | $Al_2O_3$ abrasive content (wt %) | $C_{Khphthalate|inlet}$ (molar) | RR (nm/min) | Comments |
|---|---|---|---|---|---|---|
| 2.5 | 30 | 50 | 0.0 | 0.0 03 | 0.4 | negligible film removal; film appears unpolished |
| 2.5 | 30 | 50 | 0.0 | 0.012 | 4 | negligible film removal; film appears unpolished |
| 2.5 | 30 | 50 | 0.0 | 0.024 | 14 | measurable film removal; film appears unpolished |

*Table C.1* (continued)

| P (psi) | V (rpm) | $d_p$ (nm) | $Al_2O_3$ abrasive content (wt %) | $C_{Khphthalate|inlet}$ (molar) | RR (nm/min) | Comments |
|---|---|---|---|---|---|---|
| 5.5 | 30 | 50 | 1.0 | 0.012 | 279 | borderline film adhesion; onset of heavy scratching |
| 5.5 | 45 | 50 | 1.0 | 0.012 | 355 | borderline film adhesion; onset of heavy scratching |
| 5.5 | 60 | 50 | 1.0 | 0.012 | 450 | Borderline film adhesion; onset of heavy scratching |
| 2.5 | 30 | 50 | 3.0 | 0.000 | 111 | film hazy; covered with fine scratches; no adhesive fails |
| 2.5 | 30 | 50 | 3.0 | 0.003 | 105 | film hazy; covered with fine scratches; no adhesive fails |
| 2.5 | 30 | 50 | 3.0 | 0.012 | 96 | film hazy; covered with fine scratches; no adhesive fails |
| 2.5 | 30 | 50 | 3.0 | 0.024 | 112 | film hazy; covered with fine scratches; no adhesive fails |
| 2.5 | 30 | 100 | 1.0 | 0.000 | 474 | heavy film scratching; loss of adhesion |
| 2.5 | 30 | 100 | 1.0 | 0.006 | 458 | heavy film scratching; loss of adhesion |
| 2.5 | 30 | 300 | 1.0 | 0.000 | >1000 | Complete film adhesive and cohesive failure |
| 2.5 | 30 | 300 | 1.0 | 0.006 | 684 | heavy film scratching; loss of adhesion |

This data is presented for completeness and indicates that changes in CMP process conditions can impair significantly the removal rate and post-CMP surface conditions achievable with an optimized process.

# INDEX

aluminum (Al) metallization, 1, 3, 53, 186
angle-resolved surface analysis, 78-84, 107-109, 216
atomic force microscopy (AFM), 74-77, 93, 103-106, 212-213

back-end-of-the-line (BEOL), 1, 7, 185, 191-195
benzocyclobutene (BCB),
    CMP, 58, 71-74, 86-87, 91-93, 209
    hardness, 87-90
    processing, 31, 203-205
    properties, 10, 31-31, 39
    structure, 28-30, 78-83, 221-223
    surface roughness, 74-77
    wafer bonding, 194-196

chemical-mechanical planarization (CMP),
    abrasive-free, 62, 183
    copper CMP, 50-52
    damascene patterning, 52-55
    fixed abrasive pad, 183
    FLARE, 60-61, 145, 157
    FSG, 56
    process overview, 47-49
    oxide, 50
    low-$\kappa$, 12-13, 55
    polyimide, 60
    process models, 13, 62-67

    BCB, 58-60, 71-74, 86-87, 91-93, 209
    SiLK, 61-62, 73-74, 87-88, 91-93, 98, 101-102, 225-226
    silsesquioxanes, 57
    OSGs, 56-57, 97-102
copper (Cu) metallization,
    damascene patterning, 1, 4, 52-55, 110-116
    liners, 12

damascene patterning,
    3-D integration, 193-197
    dual (DD), 1, 12, 53-54, 184
    OSGs, 110-116
    SiLK/copper, 128
    Ultra low-$\kappa$, 182

electrochemical mechanical deposition (ECMD), 184
electropolishing, 184
ellipsometry, 99, 211-212

FLARE,
    CMP, 60-61, 145, 157
    properties, 29-30
    wafer bonding, 194-198
fluorinated hydrocarbons, 34-35
Fourier-transform infrared spectroscopy (FTIR), 106-110, 217-218
front-end-of-the-line (FEOL), 1, 7

hydrogen silsesquioxane (HSQ),
    properties, 21-24
    CMP, 57-58
    wafer bonding, 194

integrated circuits (ICs), 1, 35, 185-188
interconnect,
    alternatives, 185-193
    inter-wafer, 197-198
    performance, 1, 2-14
    technology, 2-7, 86, 181-185
interlevel dielectrics (ILDs),
    elastic modulus, 37
    fluorinated glasses, 8, 19-21
    low-κ, 1-11, 20, 27-34,
    organosilicate glasses (OSGs), 9, 24-27, 97, 205-206
    oxide, 3-7, 12-13, 19-20, 50
    polymers, 4, 7-11, 20, 27-34, 58-62
    porous media, 11, 35-40
    spin-on-glasses (SOGs), 8, 21, 56
    requirements, 31
    thermal-conductivity, 40
International Technology Roadmap for Semiconductors (ITRS), 8, 11, 13-14, 182, 186

Langmuir-Hinshelwood Surface Kinetics, 13, 62-67, 119, 129, 134-138, 152, 155

methyl silsesquioxane (MSQ),
    properties, 21-24
    CMP, 57-58
    porous, 38
minimum feature size (MFS), 2, 6, 7, 13, 187-188

multiplexed interconnects, 185-188

nanoindentation, 88-90, 213-214

National Technology Roadmap for Semiconductors (NTRS), 6, 14

optical interconnects, 7, 181, 187-188
organosilicate glasses (OSGs),
    chemistry, 9, 24-27, 109-112
    CMP, 57, 91-102
    damascene patterning, 110-116
    properties, 24-27
    scratches, 105-106
    surface roughness, 103-106
    processing, 205-206

parylene,
    CMP, 58-60
    properties, 10, 33-34
    structure, 28, 33

phthalic acid and potassium hydrogen phthalate, 115, 122-129, 152-156
planarization alternatives, 181-185
polyimide
    properties, 10 33-34
    CMP, 58-60

R-C delay, 1, 14

Semiconductor Industry Association (SIA), 6
SiLK,
    CMP, 61-62, 73-74, 88, 91-93, 98, 101-102, 225-226
    hardness, 87-90
    processing, 225-226
    properties, 11, 30-34
    structure, 10, 28
    surface roughness, 74-77
slurry,
    particle size distribution, 125-127
    phthalate-based, 115, 127-129, 152-156

QCTT1010, 72-73, 120-124,
    152-154, 206-208
  potassium dichromate, 156, 171,
    177-179
  colloidal silica, 208-209
spin-etch planarization, 183-184
sputtered depth profiling, 84-85,
    221-224
surfactants, 58-60, 91-93, 206-208

thermal conductivity, 10, 35, 39-40
three-dimensional (3D) ICs, 185,
    188-199
tungsten (W) vias, 1, 5

wafer bonding, 190-199

X-ray photoelectron spectroscopy
    (XPS), 21, 62, 78-85, 106-109,
    221-224

Printed in the United States
68950LVS00001B/1-117